A Natural History of Beer

A Natural
History *of*
Beer

ROB DESALLE &

IAN TATTERSALL

Illustrated by Patricia J. Wynne

Yale UNIVERSITY PRESS/NEW HAVEN & LONDON

Published with assistance from the Louis Stern Memorial Fund.

Yale University Press books may be purchased in quantity for
educational, business, or promotional use. For information, please
e-mail sales.press@yale.edu (U.S. office) or sales@yaleup.co.uk
(U.K. office).
Designed by Mary Valencia.
Set in Baskerville and Avenir type by
Tseng Information Systems, Inc.
Printed in the United States of America.

Library of Congress Control Number: 2018951186
ISBN 978-0-300-23367-4 (hardcover : alk. paper)
A catalogue record for this book is available
from the British Library.
This paper meets the requirements of ANSI/NISO Z39.48-1992
(Permanence of Paper).
10 9 8 7 6 5 4 3 2 1

Illustration on frontispiece and title pages: iStock/micropic

To Erin and Jeanne,
even though they prefer wine

Contents

Preface

Beer is possibly the world's oldest alcoholic beverage, and it is certainly the most important historically. What is more, although beer has tended to lag behind wine in public esteem, in its more inspired manifestations it has at least as much to offer as wine does to our five senses, and to our human capacity for aesthetic appreciation. Indeed, it has been argued that beer not only is both conceptually and operationally more complex than its rival, but also can offer a more complete expression of its makers' intentions. None of this means, of course, that we lack enthusiasm for wine — as we hope any reader of our book *A Natural History of Wine* will immediately understand. Wine occupies a unique and important place in human experience, and in our own lives. But then again, so does beer; and it is clear to us that the two beverages, while complementary, are wholly distinct. If one of them merits consideration from a natural history perspective, so does the other.

Hence this book, which appears in a golden age for beer drinkers virtually everywhere. True, the recent excitement in craft brewing has unfolded against a monolithic backdrop of rather uniform mass-market beers, produced and sold in mind-numbing quantities by international giants. But at the more innovative end of the market, beers have never been produced in such variety and with such amazing inventiveness. The abundance of creative new offerings has had the effect of making the world of beer not only an exciting place but a rather confusing one as well, with an almost incomprehensible riot of consumer choice available through an archaic distribution system that makes many well-reputed beers hard to find. But sometimes a bit of anarchy can be exhilarating.

There are plenty of publications that will help you navigate the chaos, though frankly the craft is developing so fast that it is a full-time job just to keep up. Our intention here, though, is very different. Our goal is to show just how complex the identity of beer is, by situating it first in its historical and cultural contexts, and then in the setting of the natural world from which both its ingredients and the human beings who make and drink it have emerged. In the process, we traverse evolution, ecology, history, primatology, physiology, neurobiology, chemistry, and even a bit of physics, in the hope of offering a more complete appreciation of the wonderful pale-straw-to-blackish-brown liquid that reposes in the glass in front of you. We hope you will find the journey as enlightening as we did.

This book was enormous fun to write, and even more fun to research. For help with the latter, we must thank many good friends and colleagues. Among them we wish particularly to mention Heinz Arndt, Mike Bates, Günter Bräuer, Annis Cordy, Mike Daflos, Patrick Gannon, Marty Gomberg, Sheridan Hewson-Smith and the University Club of New York City, Chris Kroes, Mike Lemke (who originally taught RD to homebrew, two decades ago), George McGlynn, Patrick McGovern, Michi Michael, Christian Roos, Bernardo Schierwater, and John Trosky. We also want to express our appreciation to our favorite drinking establishments in New York City. There are many, but among them ABC Beer Company, The Beer Shop, Carmine Street Beers, and Zum Schneider come particularly to mind, just as the old West 72nd Street Blarney Castle and its incomparable host Tom Crowe remain a fond memory.

At this point in our careers we can hardly imagine producing a book without both the art and the moral support of Patricia Wynne, who is always at least as much collaborator as illustrator. Thank you, Patricia, for the pleasure of working with you, both on this project and over the years.

At Yale University Press we are above all indebted to our long-time and frequently long-suffering editor Jean Thomson Black, without whose energy, encouragement, and enthusiastic support this book would never have gone forward. We would also like to express our gratitude to Michael Deneen, Margaret Otzel, and Kristy Leonard for help with production and contract matters, Julie Carlson for her excellent copyediting skills, and Mary Valencia for the book's elegant design.

Finally our thanks are due, as ever, to Erin DeSalle and Jeanne Kelly for their patience, forbearance, and good humor at all stages of the book's gestation.

PART ONE

Grains and Yeast

A MASHUP FOR THE AGES

1

Beer, Nature,
and People

If a howler monkey could get happily sloshed, then so could we. "White Monkey," said the label on the tall bottle, the eponymous primate having apparently presided, hands across eyes, over the three-months-long aging of this Belgian-style tripel in white wine barrels. Eyes open, we unwound the wire cage, popped out the Champagne-like cork, and admired the bubbles lazily rising through the golden-amber ale. Those wine barrels were subtly detectable on the nose, but the beer hit the palate as a classic harmonious tripel, with sweet malty tones and a decadent finish. We hoped the original drunken howler monkey had enjoyed his fermented *Astrocaryum* fruits half as much!

H uman beings may be the only creatures who make beer. But if we take a broad view of what "beer" is, they are not the only creatures who drink it. As any thirsty paleontologist who has scoured the torrid Arabian landscape with only a pallid "near beer" to look forward to at the end of the day will tell you, the key ingredient of this marvelous beverage is ethyl alcohol. Yet there is nothing intrinsically amazing about this simple molecule, which turns out to be astonishingly widely distributed in nature. Vast clouds of it, for example, swirl around the center of our Milky Way galaxy, giving rise to what our colleague Neil deGrasse Tyson has called "the Milky Way Bar." Far surpassing anything in the famous bar in the first *Star Wars* movie, the alcohol molecules in this galactic cloud add up, by Tyson's calculation, to something like "100 octillion liters of 200-proof hooch." Disappointingly, though, the alcohol molecules offered by the Milky Way Bar are so vastly outnumbered by those of water that, in combination, they would yield a beverage of only 0.001 degrees proof.

Better, then, to look a bit closer to home. Although the numbers may be less extravagant here on Earth, the results are a lot more interesting. As we explain in Chapter 8, the yeasts that convert sugars into alcohol are omnipresent in the environment, just waiting for the raw materials to become available. And there is a lot of sugar around in the global ecosystem for those yeasts to work on, especially since, toward the end of the Age of Dinosaurs, some plants began producing flowers and fruit to attract pollinators and seed dispersers. The Bertam palm of Malaysia, for example, bears large flowers that exude a sugar-rich nectar. This nectar spontaneously ferments to produce a pungent beverage with some 3.8 percent alcohol by volume (ABV), which is about the strength of the beer traditionally served in British pubs.

This generous provision has been noticed by a whole variety of forest residents, but it is most particularly loved by our very remote relative the pen-tailed tree shrew. During the flowering season these tiny (chipmunk-sized) creatures binge on fermenting Bertam nectar for hours at a time. In a single session, a pen-tail might consume an amount equivalent to two six-packs of beer for an adult human, yet it will do so without showing any signs of inebriation. That is just as well, because the tree shrews' habitat is rife with predators, and even a momentary slowing of their reflexes could well prove fatal. Nobody knows how the tree shrews pull off this remarkable trick; but what is clear is that the palm nectar's appeal to the tiny mammal goes far beyond the merely nutritive.

A similar attraction to the products of natural fermentation is shown by our closer relative, the howler monkey of South and Central America—which, unlike the tree shrew, seems to feel the buzz. Back in the 1990s, primatologists studying howlers in Panama noticed one individual feeding with unusual enthusiasm on the fruits of the *Astrocaryum* palm. So frenzied did the monkey become that the observers suspected he might be drunk; analysis of the alcohol content of fruit that he had let drop to the forest floor confirmed that he almost certainly was. By the researchers' rough calculations, the twenty-pound howler had consumed the human equivalent of ten bar drinks in a single session.

This and other observations led the biologist Robert Dudley to wonder about the origins of the widespread (though far from universal) fondness among living creatures for naturally fermented alcohol. He eventually concluded that the main importance of alcohol to primates lies in the signal it carries from the plant (which wants its seeds ingested, and thereby ultimately spread around the forest) announcing the presence of the sugars being fermented. Fermentation sends out strong fumes, guiding keen-nosed fruit-eaters toward all that nutritious ripe fruit, and giving them a clear dietary advantage. The logic here applies even to the evolution of humans, because although our species *Homo sapiens* is famously omnivorous today, there are good reasons for believing that we are descended from a primarily fruit-eating ancestor.

If Dudley's "drunken monkey" hypothesis is correct (and not everybody buys it), we can view our own human predilection for alcohol as, ahem, an "evolutionary hangover." As such, this tendency of ours was probably inconsequential for as long as the only alcohol around was the little bit that mother nature spontaneously produced. It is only very recently—and, in evolutionary terms, entirely accidentally—that things have gotten a bit out of hand with the development of technologies to produce limitless amounts of alcohol at will.

Still, if we examine the matter more closely, it begins to look a bit more complicated than the drunken monkey explanation suggests. For a start, alcohol and many of its derivatives are toxic to many organisms, including most primates. Indeed, the ancestors of today's yeasts are believed to have hit on producing alcohol specifically as a weapon against the other microorganisms with which they jostled for ecological space. And although it has certainly given them a major edge in this regard, in sufficient concentrations (usually around 15 percent ABV in wine, less in a beer), alcohol is also toxic to the yeasts themselves. This is not a significant issue out there in the natural world, but it is a very important consideration in brew houses and wineries.

Closer to home, an unfortunate hedgehog was reported to have expired after lapping up a lot less egg liqueur than it would have taken to make him legally drunk in New York State. Even more suggestively, at least as many mammalian fruit-eaters (primates included) are said to be repelled by alcohol fumes as are attracted by them. Plainly, there is something a little unusual about being drawn to alcohol, and something even more unusual about being able to handle relatively large quantities of it—as we humans are to a certain extent, and tree shrews are in spades.

So where does our (modestly impressive) human tolerance for alcohol come from? As we discuss in more detail in Chapter 13, our physiological ability to handle beer and other alcoholic beverages derives from our bodies' production of a class of enzymes called alcohol dehydrogenases. Manufactured in a variety of internal organs, these enzymes break down alcohol molecules of all sorts into inoffensive smaller components. One kind of alcohol dehydrogenase, called ADH4, is present in the tissues of the tongue as well as in the esophagus and stomach, and is thus the first molecule of its kind your beer encounters when you imbibe it. Like other alcohol dehydrogenases, ADH4 is far from monolithic; instead it comes in a whole host of different versions. Some of these go straight for the ethyl alcohol molecules; others attack different alcohols, as well as the terpenoids that are widely present in plant leaves, an important source of sustenance for many of our primate relatives.

Molecular biologists have compared the distribution of alcohol-active ADH4s among a broad sampling of primates ranging from bush babies, to monkeys, to chimpanzees and humans. In doing so they discovered that, some ten million years ago, there was a dramatic switch in the pre-human lineage from "ethanol inactive" ADH4 to an "ethanol active" form. Caused by a single gene mutation, the change to the ethanol active version of the enzyme resulted in a forty-fold increase in the body's ability to metabolize ethanol.

Just why this transition occurred is hard to say. Indeed, it might well have been an adaptively random event rather than one associated with a specific dietary shift. Researchers trying to make a causal link have suggested that the relatively large-bodied primate that first acquired the enzymatic innovation might have spent increasing amounts of time on the forest floor, precisely where it was most likely to encounter the ripest and most actively fermenting fallen fruit. But since fermenting fruit can only make up a fraction of the diet of even the most dedicated frugivore, it is doubtful that a more efficient use of this resource could have alone accounted for this physiological innovation.

What is more, while the fateful shift certainly occurred in an ancient human precursor, the ancestor concerned appears to have lived *before* the evolutionary split between humans and their closest relatives, the chimpanzees and gorillas. It thus also existed before our ancestors had become the omnivores that humans are today—which, in turn, means that the change was not associated with anything that humans or their close extinct relatives uniquely do, or did. Still, whatever the initial context of this notable physiological innovation may have been, it certainly preadapted more recent humans to handle ethyl alcohol after they had (very much later) figured out how to make it in significant quantities.

This doesn't mean, of course, that early hominids (early members of our own lineage) may not have had—and even joyfully expressed— a predilection for the alcohol that mother nature, whatever her motives, had generously given them the ability to tolerate. It is far from uncommon for naturally occurring sugars (in honey, nectar, or fruits) to ferment spontaneously into alcohol; and despite the aversion to alcohol and its fumes that is also reported among fruit-eating and other organisms, the literature is replete with anecdotal accounts of animals of many kinds—elephants, moose, cedar waxwings, those howler monkeys—getting happily hammered on overripe fermenting fruits. It is hard to imagine that our early ancestors did not at least occasionally indulge themselves in this way—and indeed, there is now a scientific account of our similarly alcohol-tolerant chimpanzee relatives doing something similar.

Researchers at Bossou in the west African country of Guinea have reported that wild chimpanzees repeatedly returned to a plantation of raffia palms that workers had tapped to obtain sugar-rich sap. Dripping into plastic containers, the sap would rapidly and spontaneously ferment into prized palm toddy, which was normally collected by the laborers at the end of the day. But the workers had other duties, and while their attentions were elsewhere the chimpanzees would illicitly retrieve the toddy, crumpling leaves to form "sponges" that could be dipped into the full plastic containers. They would then eagerly suck the liquid from the loaded sponges. The researchers estimated that, at the point at which it was consumed by the apes, the toddy had typically reached a pretty respectable ethanol content of 3.1 percent ABV. Sometimes it was as high as 6.9 percent.

Palm toddy begins fermentation as a sweet and delicate-tasting beverage. But by the time the alcohol content has risen to the higher levels achieved at Bossou, the liquid invariably becomes pungent and—to us—quite repellent. Nonetheless, the chimpanzees appear to have loved it, dipping and emptying their sponges on average almost ten times a minute, for minutes on end. And while the sugar-rich sap is undoubtedly nutritious, there can be little doubt that the chimpanzees also greatly appreciated the alcoholic buzz that came along with it. The researchers certainly noted "behavioural signs of inebriation" among some of their subjects, though they reported no actual rowdiness. At least at Bossou, chimpanzees apparently don't seriously overindulge, though some went right to sleep after finishing their drinking sessions.

Still, while apes may enjoy the buzz they get from alcohol—and early human precursors almost certainly appreciated it too—there is an added dimension to our own modern human experience of the ethanol molecule. This is because only *Homo sapiens*, as far as we know, possesses the kind of cognition that not only allows us to predict the future consequences of our actions, but also gives us an awareness of our impending mortality. This knowledge places an existential burden on humankind that no other species faces: a burden that, of all available drugs, alcohol most benevolently helps alleviate.

Members of our species are uniquely capable of worrying not only about what is happening to them right now, but about things that *might* happen to them in the future. And since we know that our lives are uncertain and fraught with hazard, we welcome anything that will help distance us from this unpleasant reality. Through its inebriating effects, alcohol helps us to keep that distance; and beer delivers that alcohol both pleasantly and sociably. The French gastronome Jean Anthelme Brillat-Savarin had this all figured out almost two centuries ago, when he wrote that two significant characteristics differentiate us from the beasts: fear of the future, and desire for fermented liquors. As an added attraction, our peculiarly human cognitive style also allows us to process the input from our senses in an entirely unprecedented way, making it possible for us to analyze our experience of what we are drinking in aesthetic terms (see Chapter 11). This adds yet another dimension to our experience of beer, a beverage that offers us a huge variety of sensory experiences to appreciate—and to argue about.

We will delve further into mild inebriation and its charms in Chapter 13. But before we forget the important point that any fermented beverage may be quite nutritious as well as intoxicating, we might also mention that beer has occupied a very special place as a dietary resource throughout the history of sedentary *Homo sapiens*. Intimately connected both historically and chemically to bread, "the staff of life," beer is often referred to as "liquid bread." Indeed, the two substances are so closely related—after all, they are often fermented from the very same cereals by the very same yeast species, *Saccharomyces cerevisiae*—that it is still vociferously argued which came first, bread or beer.

It is probably wise to avoid that controversy here, but we do want to clear up a question that often comes up in barroom arguments involving the bread-and-beer issue. As we will discuss in some detail in Chapter 10, the byproducts of fermentation using yeasts are ethanol and carbon dioxide. When a baker makes bread, he or she mixes the dough and puts it in the oven. As the mixture heats up the yeast go to work, producing carbon dioxide gas that bubbles into the dough, causing it to rise. But what happens to the ethanol inevitably generated at the same time? Why don't we get drunk eating bread as well as drinking beer? The answer lies in the high temperatures involved in baking, which cause most of the alcohol to evaporate. But not quite all. When the bread leaves the oven, it contains a bit of residual ethanol; and while the lingering quantity is mostly minute, sometimes as low as 0.04 percent ABV, for a moment or two it might be as much as 1.9 percent ABV. No wonder fresh-baked bread smells so good! Interestingly, the 1.9 percent figure is very close to the 2 percent ABV that is the most you can metabolize as fast as you take it in. So while some bread hot out of the oven might momentarily contain half as much alcohol as the average English ale, you could never eat it quickly enough to get even slightly tipsy.

Why have humans so eagerly co-opted the natural process of fermentation? Until the end of the Ice Ages, some ten thousand years ago, all members of *Homo sapiens* were hunters and gatherers, living itinerant lives and living off whatever bounty nature provided—which would have varied greatly from place to place. There is some evidence that our hunting-gathering forebears occasionally consumed cereals, but grains did not come into their own as a major human dietary resource until climates ameliorated at the end of the last Ice Age. The rise in temperatures caused huge changes in the plant and animal resources locally available to the human populations that by this point were scattered all over the habitable world. In response to this major environmental challenge, people—independently in several different localities around the globe—adopted settled lifestyles that depended on the domestication of plants and animals.

This fateful transition to sedentary life was not a simple process, and it unfurled differently and at varying paces from place to place. But although it turned out to be a hugely Faustian bargain—hunter-gatherers are typically much healthier and more egalitarian than sedentary folks are, with much more leisure time—the time had evidently come for the new economic style. And everywhere the change happened, domesticated cereals were at the fore—wheat and barley in the Near East, rice in eastern Asia, maize in the New World.

If you are a hunter-gatherer, your economic strategy is relatively straightforward. You make use of what nature offers, something that often obliges you to move across hundreds of miles of territory in any given year. But for a settled agriculturist, growing seasonal crops at a particular place, life is more complicated. You find yourself with an embarrassment of riches in some seasons, and nothing at all to harvest in others. You thus need ways of storing food, so that nourishment will be available to you and your family year-round. Especially in the warmish places in which agriculture initially developed, however, preserving stored food can be a headache. Grain kept in piles or pits rapidly rots through oxidation, and even presents a risk of spontaneous combustion. Just as important, it also needs to be kept safe from the nibblings of hungry animals, ranging from hordes of tiny insects to voracious rodents.

Enter fermentation. The researcher Douglas Levey has proposed that, from an anthropological perspective, the deliberate fermentation of grain is best regarded as a sort of controlled spoilage. Most of the microbes that are responsible for the decomposition of stored food are unable to live in the presence of alcohol—it is, after all, a famous disinfectant—so by permitting a certain degree of fermentation of their grains by naturally occurring yeasts, the early farmers could preserve a lot of their nutritional value, even if not their freshness. This was so hugely important for them that Levey reckons fermentation was used as a preservation strategy before it was co-opted for making intoxicating beverages. Since the byproduct of fermentation is alcohol, and you can't have the one without the other, maybe the issue is moot. But there can be no question that, in addition to its mind-altering qualities, beer was an important source of stored nutrition in the ancient world—as indeed it was, until not too long ago, in ours.

It is worth noting that, although wine more or less makes itself as the sugars naturally present in the grape are fermented through the action of yeast, beer requires more fiddling. The cereal grains used to make beer contain long molecules of starch that have to be broken down into simpler sugar molecules before fermentation can even begin. The modern brewer's preferred way of achieving this conversion is by malting the grain—that is, soaking and aerating it to stimulate sprouting, then drying it to stop the sprouting process before the resulting sugars are consumed. The dormant sugars can then be exposed to the tender mercies of the yeast whenever required.

We cannot conclude this chapter without mentioning what some consider to be the best news about ethyl alcohol: in limited concentrations, it can actually be good for you. This proposition has been tested on fruit flies, a favorite subject of laboratory researchers because they are easy to keep, and reproduce very rapidly. And it turns out that fruit flies exposed to moderate concentrations of alcohol fumes live longer and reproduce more successfully than either "heavy drinkers" or "teetotalers." What is more, larval fruit flies infested with parasites have been observed medicating themselves by preferentially seeking out foods containing ethanol. Perhaps less promisingly, adult flies prevented from mating showed an increased attraction to ethanol, perhaps as if to drown their sorrows.

In humans, clinical studies have repeatedly associated light-to-moderate drinking with a reduced prevalence of a whole spectrum of individual diseases, and with an overall decreased risk of death. The cardiovascular system seems to benefit especially here, with moderate alcohol consumption being solidly associated with such advantages as lowered hypertension, reduced LDL and higher HDL cholesterol levels, and a diminished probability of ischemic stroke. A 2017 study that tracked more than 300,000 people over an average of about eight years found that, compared to lifetime abstainers, light to moderate drinkers were about 20 percent less likely to die of any cause over the tracking period—and were 25 to 30 percent less likely to die from cardiovascular disease. Among other specific conditions, lowered incidences of diabetes and gallstones have also been reported among moderate drinkers, though on the other side of the ledger, recent studies have also raised the possibility of an association between moderate drinking and breast cancer in young women.

Overall the indications seem to be that, for most of us, the health benefits of moderate drinking considerably outweigh its risks. But moderation seems to be key, since there is no questioning the health and social damages that are wrought by excessively high intakes of alcohol— damages that overwhelm any benefit that the alcohol might or might not confer at lower levels of consumption. That same 2017 study found that over the period it covered, male heavy drinkers had a 25 percent increased risk of mortality from all causes compared to the lifetime abstainers, and a whopping 67 percent additional risk of dying from cancer. As we stress in Chapters 12 and 13, quite aside from the terrible social effects of alcoholism, these figures on their own make a compelling argument for avoiding the immoderate consumption of any alcoholic beverage. Still, in a perverse way they are (relatively) good news for beer drinkers, whose preferred beverage has an inherent advantage over drinks that deliver more alcohol per sip.

2

Beer in the Ancient World

This, the parent of all modern ales, was one we had to make ourselves. Into the pot went New York City water, a frighteningly large amount of fresh-milled two-row barley, and a sockful of flaked barley for good measure. After boiling this mixture, we added a gruit of hibiscus and other herbal and citrusy ingredients, then let it ferment with available yeast. A month later we siphoned the dark brown liquid into bottles, then aged it for two long weeks. Our first bottle opened with a satisfying hiss. Now soupy and yellowish-brown in color, our simple gruited ale was pleasingly sour on the palate, with a grassy aftertaste. We were very agreeably surprised at how drinkable it was. No wonder those Iron Age Germanic tribes had clung so tenaciously to their brewing traditions.

The first literary mention of beer firmly positions the beverage as a civilizing influence. In the epic poem *Gilgamesh,* the mythologized account of a Sumerian king who reigned around 4,700 years ago, a wild man called Enkidu is brought into a village and entreated to "Drink beer, as is the custom of the land." Only once he had drunk the beer, and consumed the bread he was also offered, was the feral Enkidu at last considered ready to enter civilized society, and to proceed to Gilgamesh's capital city of Uruk. And what could have been more emblematic of civilization than beer and bread? For the very existence of the fabled metropolis of Uruk had been made possible by the productivity of the vast and astonishingly fertile cereal-growing plain that lay between the Tigris and Euphrates rivers in Mesopotamia, the "land between two streams." The Sumerian Empire, like the Babylonian one that followed it, was built on grain, and on the beer and bread made from it.

By Gilgamesh's time, settled life and the raising of the cereals from which beer is made already had a respectably long history. As noted earlier, the ancestral human hunting-gathering way of life began to be abandoned when the huge polar icecaps started to shrink due to climatic warming at the end of the last Ice Age. In their wanderings around the landscape, ancient hunter-gatherers must have occasionally encountered naturally fermenting fruits and honey, and the alcohol they produced. Although it is doubtful that any fully itinerant human societies ever possessed the technology necessary to malt and ferment significant quantities of cereals, it is unlikely that, with the advent of settled life, much time was allowed to pass before brewing began on a significant scale.

In the Near East, where wheat and barley were first cultivated, the transition from itinerant to settled life is particularly well documented at a Syrian site called Abu Hureyra. Between about 11,500 and 11,000 years ago, the people who camped there were still practicing a traditional hunting and gathering lifestyle. By around 10,400 years ago, their descendants had begun to supplement their diets with cultivated cereals; and by 9,000 years ago, residents' food supply came principally from domesticated animals and plants of various kinds—although plenty of wild gazelles continued to be slaughtered on their annual migrations through the area.

Over this period, Abu Hureyra itself developed from a gaggle of simple roofed-over "pit dwellings" excavated in the ground, to a substantial village of clustered mud-brick houses and open courtyards. Rather unusually, the Abu Hureyra villagers first selected rye for cultivation; in the Near East more generally, barley as well as einkorn and emmer wheats were the cereals of choice for domestication, making this a leading region for the invention of barley-based beers.

Interestingly, the domestication of cereals occurred before the invention of pottery, which first turns up in the Near East about 8,200 years ago. Although pottery might not have been entirely essential for making beer in some form, it was certainly a prerequisite for making it in any quantity. By the time it came along, people had been grinding cereals for a long time, with an early example of cereal grinding from some 23,000 years ago giving us one reason for suspecting that bread may have preceded beer in the human diet. What is more, large, hollowed-out stone containers as much as 11,600 years old at the pre-Neolithic site of Göbekli Tepe, in eastern Turkey, may have contained a beverage fermented from wild cereals.

When pottery vessels first came into use, settlements were small, and people lived in relatively egalitarian communities of a few hundred at most. Most members of those communities were related to one another, worked in the fields together, and shared similar skills. But change happened fast. By five thousand years ago, around the time that the newly civilized Enkidu was spirited off to Uruk, a strongly stratified society had already developed in Mesopotamia. Specialized skills had proliferated, and social roles and status had become strongly differentiated. Most citizens still labored in the fields, but the more important of them lived in towns and in the burgeoning new cities. Some of them were brewers. And those early brewers, it seems, were women.

No one knows exactly when people started brewing in the newly available ceramic pots. The earliest chemical traces of barley beer, discovered in the form of calcium oxalate (beerstone) deposits in a pottery jar from the Sumerian outpost of Godin Tepe, in northern Iran, date from only a little over five thousand years ago, making the liquid from which the beerstone had precipitated more or less contemporary with our friend Enkidu. But nobody would doubt that the Near Eastern brewing tradition is much more venerable than this; we wouldn't be surprised if beerstone were eventually to turn up in one of the very earliest pottery vessels from the region.

If we don't know exactly how far back the brewing tradition goes in Mesopotamia, do we at least know what the product was like? To our great good fortune, the answer is a qualified yes. For some clay tablets bear an inscription that has become known as the *Hymn to Ninkasi*, and Ninkasi was the Sumerian goddess of beer. Happily, the *Hymn* is not only a paean to the goddess herself, but also gives us a (sort of) recipe for a brew that was presumably produced by her priestesses, and that must broadly have resembled what women brewed at home for their families. This recipe was clearly only one of many, for the Sumerians recognized at least twenty different types of beer: white, red, black, sweet, "of superior quality," and so forth, frequently flavored with exotic aromatics.

Ninkasi's beer was probably quite different from the brew you may have sampled on your way home from work today. In the *Hymn*, Ninkasi is described not only as preparing the malted (sprouted) grain by soaking it in water, drying it out to arrest the sprouting, and "brewing it with honey [possibly better translated as 'date juice'] and wine," but also as baking bappir, a barley bread, presumably as a vehicle for getting yeast into the beer. But whether or not the bread played that role, in the end Ninkasi was obliged to pour the probably violently fermenting final product into a collector vat before serving it like "the onrush of the Tigris and Euphrates."

However it was produced, Ninkasi's brew is generally believed to have been a rather soupy and cloudy product—and certainly, abundant floating solids might explain why it was typically drunk from a large common jar, often the fermenting vessel itself, using long drinking straws. Once served, it seems to have been rapturously received, for the *Hymn* records that it "rejoices the heart." This is a sentiment with which all modern beer-lovers will readily agree, although their doctors might frown a little at the poet's other claim, that beer "makes the liver happy."

Chapter 15 considers the experiences of various brave souls who have tried to re-create ancient beers such as Ninkasi's. For the moment, we will just note that, whatever its other properties might have been, with its added date juice or honey and wine Ninkasi's beer (of which one modern replica came in at a respectable 3.5 percent ABV) definitely qualifies as an "extreme" beer of the kind enjoying a renaissance today among more adventurous beer lovers. Clearly, beer did not start out as a simple beverage that became more complex over time. Indeed, it would seem more accurate to say that today's craze for extreme beers represents a return to the beverage's origins.

One difference between beer in earlier times and beer today is that nowadays we have the option of slaking our thirsts with water. Today most citizens of developed economies take pure and refreshing water for granted; but it was not always so. The Agricultural Revolution brought with it pollution on a grand scale, as one of the major byproducts of progress with which human society is still coming to terms. On the marshy Mesopotamian plain, crowded with people and their even more numerous domestic animals, there would have been few sources of reliably potable water in Sumerian times. This meant that if you couldn't afford the wine available only to a privileged few, the safest option was to drink Ninkasi's brew. And thus it has been in most places, throughout most of recorded history.

Any beverage with its own goddess must have been very significant to the society that produced it, and maybe beer's purity would have been enough on its own to give it this status. But the brew's significance to the Sumerians went far beyond this, for it was a major means of distributing wealth within Mesopotamian society. Taxes were often paid in the form of grain that was presented to the temple. Priestesses of Ninkasi and other deities would then transform this grain into beer (and bread), with the results of their labors dispensed to the population as payment for services rendered. Cuneiform tablets indicate that laborers would receive one *sila* (about a liter) of beer a day. Low-level functionaries would get two, and so on up to the highest officials, who would get five *sila*. Beer in those days didn't keep long and needed to be drunk quickly. But this doesn't mean that those five-*sila* luminaries were constantly sloshed; the beer, short-lived as it probably was, was still good for smaller payments back down the line.

None of this is to say that beer was simply of economic and epidemiological significance to the Sumerians. Then as now the most social of beverages, beer also had huge symbolic importance in Sumerian society. Shared from those communal fermenting jugs by humble farm laborers and nobility alike (the plebs using simple reed straws, the nobles employing elaborate tubes of gold, bronze, lapis lazuli, or silver), it was the drink over which the various strata of Sumerian society bonded. Beer flowed even at the greatest state occasions. In 870 BCE, Ashurnasirpal II, king of Assyria, held what may rank as the most over-the-top celebration ever, to mark the completion of his new capital at Nimrud (a city south of today's Mosul and recently a major target of desecration by ISIS). In those happier days Ashurnasirpal welcomed some seventy thousand guests for ten days of feasting, during which ten thousand multi-liter jars of beer were consumed, along with the roasted carcasses of many thousands of sheep, cattle, and other unfortunate animals, and a further ten thousand skin bags of wine.

Beer, then, rather than love (Ashurnasirpal was a particularly vicious warlord, and proud of it) was what made the ancient Mesopotamian world go around. And in a sad harbinger of what was to become a standard pattern throughout history, an abundance of laws and regulations rapidly came to govern its consumption. Early in the second millennium BCE, the Babylonian king Hammurabi published a legal code regulating the conduct of his citizens, including their drinking habits. One of Hammurabi's injunctions might fall under the rubric of consumer protection, specifying that tavern-keepers (women, it appears) who shortchanged their clients should be drowned. But another is more darkly political, threatening those same tavern-keepers with death if they failed to report overheard conspiracies. Even in those early days, taverns seem already to have become places of energetic political debate and possible sedition—and probably dens of vice, to boot.

Mesopotamia may have been where barley-based beers were invented, but the ancient Egyptians were equally enthusiastic about them, and brewed a generally more sophisticated version. They, too, awarded beer its own goddess, Tenenit, although the beverage was commonly associated with a more senior goddess, Hathor, at whose temple in Dendera an inscription from around 2,200 BCE reads, "The mouth of a perfectly contented man is filled with beer."

Legend had it that the great god Osiris himself gave Egypt the gift of beer. But chances are that, very early in their history, the Egyptians actually picked up from the Sumerians the habit of brewing (and the tradition of women brewers, though men took over later). The beers of the two great early civilizations were certainly of similar nature. Usually made using a crumbled barley bread that may have included some malted (sprouted) grain, Egyptian beer was thick, nutritious, and often quite sweet, particularly when it was flavored with the dates and honey that the early clientele favored. Later, it tended to be brewed directly from barley and emmer wheat that was mixed with unroasted malt before fermentation began. Experimentation was clearly the brewers' vice right from the start; though then, as now, variation in quality was sometimes an economic issue as well as a gustatory one.

As it did for the Sumerians, beer also played an important role in the social life of the ancient Egyptians. All ages and classes drank the stuff, wages were paid in it, and it was an important feature of religious festivals. The artisans who built the great Pyramids of Giza were recompensed partly with beer, their cups replenished three times daily to a total of about four liters. The smallest and most recent of the pyramids, that of the fourth-dynasty pharaoh Menkaure, was built around 2,500 BCE by a vast crew of laborers including a gang who called themselves, according to an inconspicuous and illicit inscription, the "Drunkards of Menkaure."

We don't know how rowdy these laborers became on their tipple, but evidently beer was the lubricant that made the astonishing feat of pyramid-building possible. And if all that backbreaking labor took a toll on your health, no problem: the therapeutic benefits of beer were also widely touted. Brewed with a vast and varied array of additives, the beverage appeared as an ingredient in early Egyptian physicians' prescriptions for a whole slew of different ailments.

Once beer had put you back on your feet, it was evidently omnipresent as an adjunct to civilized life. Some of the most charming

and intimate of the many scenes of ancient Egyptian existence represented on the tomb walls of the wealthy involve making and drinking
beer—and even regurgitating it. And if, in the end, you were unfortunate enough to succumb to an affliction untreatable even by beer, the
beverage was an essential adjunct to accompany you into the afterlife.

Egyptians thus took access to their beer seriously. Queen Cleopatra VII (yes, *that* Cleopatra) provoked considerable ire among her
subjects for taxing beer (apparently a historic first) to pay for her wars
with Rome. Those beer-loving citizens were presumably even more dismayed when Rome finally won, for the Roman attitude toward the beverage was more than a little disdainful. The oenophile historian Tacitus,
for example, referred to beer as a "horrible brew" that bore "only a very
far removed similarity" to his preferred tipple. In a similar vein, the em-
peror Julian compared the scent of wine to nectar and the smell of beer
to that of a goat. Wine, in other words, came from the gods, while beer
was a coarse human product.

Given beer's dubious reputation in ancient Rome, it is remarkable that
the first recorded brewer in Britain was a colonizing Roman named
Atrectus. He and his fellow expats in the wild and woolly northern
outposts of the Roman Empire had presumably picked up the habit
of brewing from the uncouth Iron Age Germanic and Anglo-Saxon
tribes surrounding them. These reluctant new subjects of Rome were
descended from pioneering agriculturalists who had appeared in the
cold and inhospitable reaches of northern Europe much earlier, around
the time that Enkidu was luxuriating in the civilized amenities of Uruk.
They had clearly brought beer with them, because their practice of
brewing is documented amazingly early, most notably at the 3,200–
2,500 BCE Neolithic site of Skara Brae in the remote and windswept
Orkney Islands north of Scotland.

Thanks to excavations at a site in Germany dating from around
2,500 years ago, we also know a bit about how the Iron Age tribes of
northern Europe prepared their malts. Barley was apparently soaked

in specially dug ditches until it sprouted. Fires lit at the ends of the ditches then terminated germination. The resulting smoke would have imparted a dark color and a smoky taste to the malt, which was probably of remarkably high quality. If the seeds of the mildly toxic henbane plant, also discovered at the site, had been added to the brew, the final product would have been quite potent—though it would have tasted very different from your average beer today.

Figuring out how to grow cereals in the cool, rainy environments of northern Europe was no mean feat, so it is hardly surprising that the early European brewers supplemented what little barley or wheat they could spare for malting with honey, berries, and anything else fermentable they could get their hands on. Some scholars have claimed that most early European alcohol was consumed in ritual contexts, but the main evidence for this seems to be that most drinking paraphernalia have been found in tombs. Such places are certainly redolent of ritual, but they are also where such stuff is most likely to be preserved and found by archaeologists. By now, most authorities are satisfied that extreme beers were an everyday beverage in Neolithic Europe.

Beer, then, was a part of the fabric of life in northern Europe from the very start of agriculture in the subcontinent. There are hints that its consumption may not always have been as decorous as it was in Sumer and Egypt, at least according to the texts that have come down to us. The Christianized Roman observer Venantius Fortunatus, for example, once described the participants in a Germanic drinking party as "carrying on like wild men . . . a man had to consider himself lucky to come away with his life." Binge-drinking is evidently no modern invention.

Beer thus has a long—if not always reputable—history in a vast core area stretching from Mesopotamia to western Europe. But we cannot ignore that the very earliest evidence for the concoction of an at least partly cereal-based alcoholic beverage actually comes from faraway China. Beginning in the 1980s, archaeologists working in north-central China at Jiahu, the site of a Neolithic village occupied between about 9,000 and 7,600 years ago, found evidence of a remarkably sophisticated society. From the very beginning, the Jiahu people had used pottery containers; and in some of the very oldest of these, a team led by the biomolecular archaeologist Patrick McGovern discovered chemical traces of what can with some justification be described as a rice-based beer.

Technically, though, the scientists referred to the Jiahu product as a hybrid beverage, since the chemical markers that they identified almost certainly had their origins in a whole variety of ingredients. First there was that rice, possibly a short-grained domesticated variety of which whole grains were found preserved at the site. It's believed that the starches in it were broken down into fermentable sugars either by chewing and spitting (probably the earliest saccharification method devised by humans), or by the malting process later used in the West, rather than through the action of molds, as in the production of Chinese rice wine today (a later process, currently not dated earlier than the Shang dynasty in the late second millennium BCE). Then there were compounds identified variously with grapes, honey, and hawthorn fruit. Taking all this evidence together, McGovern and his colleagues concluded that the Jiahu beverage was a composite of a grape-and-hawthorn wine, a honey mead, and a rice beer. For the record, in McGovern's working definition a "wine" is fruit-based, with a relatively high alcohol by volume (ABV) of 9 to 10 percent or more, while a "beer" is cereal-based and has a lower ABV, in the 4 to 5 percent range. Significantly, though, McGovern chose to partner with a brewmaster rather than with a winemaker to re-create the hybrid Jiahu beverage; and although the result came in at 10 percent ABV, the brewery classified it as an "Ancient Ale."

Chapters 14 and 15 will look in some detail at the re-creation of this and other ancient beery beverages. For the moment, it's enough to point out that the difficulty of fitting the Jiahu concoction and other ancient alcoholic potions into any modern category underlines the fact that the first Neolithic producers of alcoholic beverages were experimenting with pretty much anything and everything that could be fermented. In those days, much as in "extreme" brewing today, anything went—although doubtless the ancient clientele soon clamored for more of the specific kinds of fermented beverages that they liked and could afford. Still, our jaunt through ancient history shows that the classification of beers we use today is a pretty recent phenomenon—or maybe even epiphenomenon.

3

Innovation and an Emerging Industry

The chilled bottle stood on the table, glistening with tiny beads of condensation. "Since 1040," said the neck label. With some reverence we levered the cap off this modern product of the world's oldest brewery. The pour was smooth, the head modest, the color a bright light amber. The flavors that followed were beautifully balanced between malt and hops. We were tasting a classic, well-crafted lager that was doubtless a far cry from the dark and cloudy ales that the monks of Weihenstephan would have made in the eleventh century. But then again, we thought, maybe something has been learned in almost a thousand years of brewing.

T he recent history of barley and wheat beers is basically European, yet beer's appeal clearly extends far beyond the subcontinent's borders. China, arguably the birthplace of cereal-based alcoholic beverages, has overtaken the United States to become the world's largest commercial beer market, consuming a mind-boggling 25 billion liters in 2016. But the making of beer in China in the modern age dates only from 1903, when Germans opened a brewery in Qingdao (Tsing Tao); and brewing there continues to be performed overwhelmingly—though not exclusively—in the German lager style. In Japan, where beer is now an ingrained part of the culture and is the alcoholic beverage most consumed, the beverage's history is only marginally longer. The first beer drunk in Japan in modern times arrived in Tokyo Bay in 1853, in the bar aboard Commodore Perry's flagship USS *Mississippi* (although Dutch traders had brewed beer there back in the seventeenth century, for their own consumption); and Japanese brewing today remains strongly influenced by the ultimately German-inspired American industrial brewing tradition. Twenty-first-century India, though hugely populous and synonymous with the world's most widely consumed style of pale ale, has been aptly described as a "beer-drinking lightweight," although reportedly sales have ticked up a bit lately. And in the Near East, where the Sumerians experimented so liberally with the possibilities inherent in fermenting cereals, the production and consumption of beer have been severely inhibited for well over a millennium by the Koranic injunction against "wine"—broadly interpreted, alas, to include all alcoholic beverages.

So back to Europe. We don't know a lot about local European brewing traditions during the probably misnamed Dark Ages, the period following the collapse of the Roman Empire in the fifth century CE. What we do know is that, while wine continued to be made and drunk in the warmer parts of the former Empire, in its cooler northern reaches, cereals reasserted themselves and beer returned to the fore. Wheat was apparently the more prestigious grain, but barley was more widely grown; and everyone, from the humblest peasant on up, drank vast amounts of usually barley-based small (weak) beer—often derived from a second usage of the malt—as a much safer alternative to water. The alcohol in the beer would have helped here, of course; but it was the boiling involved in the preparation of the beverage that assured a degree of sterility that most water sources of the time could never match. Only the rich or the aristocratic ever drank the more expensive honey-based meads and stronger beers; and, except in sacramental contexts, wine imported from the south was rarely seen in most of northern Europe. Small beer ruled the day, and beginning in about 800 CE it traveled with the Vikings on their long ships, to fortify them for their long, arduous voyages.

Just as the Romans had regarded wine as the gift of the gods, and beer as something infinitely lowlier, the emerging Christian church prized wine for its sacramental importance (the gift of just one god this time), while taking a pretty dim view of beer. In the fifth century, a minor theologian called Theodoret of Cyrrhus referred to barley beer as "vinegary, foul-smelling, and harmful." His choice of words may or may not have been an accurate description of his local brew, but it was certainly influenced by the widely held perception among Christians that beer was the drink of pagans.

In the end, though, the church authorities opted to accommodate the gustatory preferences of the people they wanted to convert. Evidently concluding that if they couldn't beat them, they should join them, the monks also found that brewing was an excellent way of using and preserving the tithed cereals with which their granaries overflowed at harvest time. And so the tradition of monastic brewing was born. Before long, monasteries were not only endearing themselves to their flocks by upping (and incidentally masculinizing) the brewing game; they were also finding that beer was a useful source of revenue. As the Middle Ages progressed, the beverage increasingly became a subject of lively monastic experimentation, as the monks brewed beer for every pocket, and—like their secular colleagues—scented their best brews with an added "gruit" of increasingly exotic herbal ingredients like burdock, yarrow, wormwood, sage, mugwort, horehound, or juniper berries.

And, eventually, hops. The ninth-century (or possibly earlier) innovation of adding the dried seed cones of the climbing plant *Humulus lupulus* to the beer-making process changed everything. Hops are not simply a powerful flavoring agent, conferring a refreshing bitterness; they also contain a natural preservative that extends the life of beer (see Chapter 9). Without hops, beer had to be drunk when it was fresh, that is, locally. Only strong beers with lots of alcohol (itself a preservative) could be shipped any distance at all. But with hops, any beer—

including less alcoholic ones that demanded less malt to brew—could be transported farther than before, allowing a relatively long-distance trade to develop.

All the early monastery beers (or abbey beers, as the living if not uninterrupted tradition is still known in Belgium) fell under the very general category of ales. Ales are beers fermented at room temperature, mainly using the yeast *Saccharomyces cerevisiae*, the same species used in baking bread and fermenting wine, but occasionally using wild yeasts (see Chapter 8). During fermentation, the yeast rises to the top of the liquid, forming a dense froth. Then, as now, the process of brewing ale lent itself to infinite variety: by adjusting a host of variables that included the temperature of the fermentation vats (in early days achieved by choosing the season for brewing), fermentation time, type and quantity of malt used, how the malt was roasted, composition of the gruit (which was gradually phased out after hops came along), amount of hopping, and identity of those hops (see Chapter 9), monastic brewmasters were able to produce a vast range of flavors, textures, and alcohol contents—although as trade grew and reputations developed, each abbey tended to specialize, at least seasonally.

The world's longest continuously operating brewery originated as a monastic enterprise. State-owned today, the German Weihenstephan brewery in the Bavarian town of Freising began beer production under the auspices of the Benedictine Weihenstephan Abbey (Figure 3.1). In 1040 CE the city granted the monks of Weihenstephan a formal license to brew beer, using hops that records show had already been grown on the property for several hundred years. It seems rather unlikely that brewing at Weihenstephan had started only in 1040, but the date on the license allows the Czech Žatec enterprise to claim to be the world's most ancient living brewery, having first paid a tax on the beer it produced in 1004. Physically, though, today's Žatec facility dates from as recently as 1801. The oldest active brewery still in monastic hands is the Bavarian Weltenburg Abbey, which began operations in 1050. Despite a

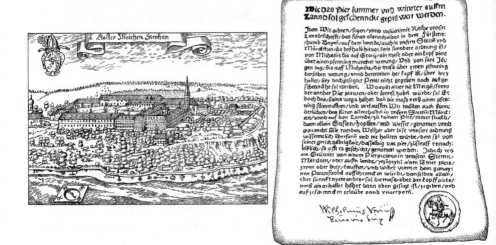

Figure 3.1. *Left:* Engraving of Weihenstephan Abbey, from Michael Wening's *Topographia Bavariae,* ca. 1700. *Right:* The 1516 Bavarian *Reinheitsgebot* decree.

brief interruption during the political upheavals of the early nineteenth century, Weltenburg is still a monastic institution, brewing an award-winning dark beer as well as a delicious Pilsner.

The monasteries' monopoly on German commercial beer-making didn't last long. With the steady expansion of German towns and cities through the late Middle Ages, the increasingly prosperous and influential commercial middle class wanted its cut—possibly one of the reasons the Freising town authorities issued that license to Weihenstephan. Brewers in the city of Cologne were finally allowed to organize a guild in 1254 (a century later than most other rising professions), and other places followed suit. This led cities and regions to develop their own styles, and to compete for trade. Inevitably, beer wars followed.

The earliest products of Weihenstephan, Žatec, and Weltenburg were all ales. Yet, with the exception of Hefeweizens, the beers produced by these historic brewing operations today are all lagers, as the result of the most momentous schism in the history of brewing. Back in the early fifteenth century (possibly earlier), brewmasters in and around Einbeck in Lower Saxony started producing a beer in a radically new style. Brewers in Bavaria were already in the habit of storing and aging their ales in limestone caves, where cool conditions inhibited the growth of undesirable bacteria as they matured. But there was something different about the new product from Einbeck. After spending the winter gently maturing in the cool caves, the beer became clear and bright, with a crisp finish — unlike the chemically more complex and typically cloudy ales of the time. In German this long-established process of cold storage is known as lagering, and nobody at the time had the slightest idea why it should uniquely have had this salutary effect in Einbeck, of all places.

This ignorance is hardly surprising, because nobody then knew just what made fermentation happen. People had long been aware that some specific element promoted fermentation, and selection for the agent concerned was implemented by transferring the floating froth that formed atop one brew to the next. But the discovery that fermentation was accomplished by the tiny living organisms we know today as yeasts had to await the research of the French chemist Louis Pasteur in the nineteenth century. Once Pasteur had made his great discovery, however, it was only a matter of time before it was realized that the brewers of what had become known as lager were making their beer with a distinctive kind of yeast (see Chapter 8). Unlike the traditional *Saccharomyces cerevisiae*, which was happiest fermenting at around 21°C, the newly recognized yeast now known as *Saccharomyces pastorianus* (for Pasteur) flourished at much cooler temperatures of around 4.5°C; and unlike the top-fermenting *Saccharomyces cerevisiae*, it descended to the bottom of the fermenting tank, taking other detritus with it and leaving the liquid above clean and bright — albeit in earlier days typically quite dark in color, because that was the result when malt was roasted in a smoky, wood-fired kiln. Just as importantly, the new yeast vented its waste carbon dioxide up through the liquid, giving it a sparkle.

Where exactly the new yeast had come from is still debated. For the last forty years it has been known that *Saccharomyces pastorianus* is the result of hybridization between *Saccharomyces cerevisiae* and another species. But the second yeast — which had evidently given *Saccharomyces pastorianus* its cold tolerance and its bottom-fermenting proclivities — long remained elusive. Now it has been identified and given the name *Saccharomyces eubayanus*. This yeast was first discovered in South America, but now it has been documented in Tibet as well. That newly broadened distribution makes it quite probable that it is also lurking, as yet undiscovered by science, in the oak forests of central Europe. If it isn't, how it found its way to Einbeck remains anyone's guess.

Meanwhile, other important developments were afoot in fifteenth-century Bavaria, the epicenter of brewing innovation. Most significantly, this was the place where the German *Reinheitsgebot* (beer purity law) originated. First promulgated in the Duchy of Munich in 1487, then codified across Bavaria in 1516 and ultimately throughout Germany (when Bavaria refused to join the Weimar Republic in 1919 unless the law was adopted nationally), the *Reinheitsgebot* decreed that the only legal ingredients in beer were water, barley, and hops (Figure 3.1). Yeast was added to the list centuries later, after Pasteur's discoveries. Significantly, the law also regulated how beer could be sold, and at what price. From one point of view this might appear to have been enlightened consumer protection legislation, but it may be more significant that barley was specified as the only cereal for beer largely because wheat shortages often placed bread in short supply. What is more, taxes on beer were a major source of revenue for the secular authorities, who were keenly aware that a decline in quality might also correlate with a decline in their finances. It was probably such concerns that prompted the Bavarian authorities in 1553 to ban brewing in the warm summer (when potentially detrimental microbes flourished), thereby more or less limiting Bavarian beer production to lager—with long-term repercussions for the beer market worldwide.

Not that German beer production is entirely monolithic. Although lager is overwhelmingly the best seller nationwide, top-fermented wheat beers are also widely made and consumed. Other German brews include rye beers; hybrid top-fermented beers such as Kölsch that are put through lagering; the famous smoke beers of Bamberg, which combine lager yeast with malt roasted over immemorial wood fires; and even a concoction from the city of Cottbus that contains honey, molasses, and oats in addition to both wheat and barley malts.

Next door to Bavaria lies Bohemia, the western part of today's Czech Republic. With a fine brewing tradition but no *Reinheitsgebot,* brewing standards in the Bohemian town of Plzeň (Pilsen) seem to have declined by the early nineteenth century—so much so that in 1838, rioting residents dumped several dozen barrels of apparently appalling local beer on the steps of its town hall. Shaken, the Pilsen city fathers appealed for help, which came in the form of the gruff and cantankerous Josef Groll. A brewer from Bavaria who had traveled to England and knew the secrets of brewing a light ale using a coke-fired pale malt (more on this later), Groll invested in a British kiln and put the resulting malt through a fermentation in the Bavarian lager style. The local soft water, Saaz hops, and barley proved ideally suited to this treatment; when the first barrels were tapped several months later, everyone was entranced. Groll's "Pilsener" was clear, light, golden and bright, with "a thick, snow-white foam" and subtle hop aromas. It set a standard for lager that others have striven mightily to match. There are now "pilsners" made all over Europe, indeed all over the world, in a variety of styles, but enthusiasts argue that only in Pilsen do all the ingredients come together to such perfection.

Although German beer drinkers were gradually abandoning ales in favor of lagers, ales continued to flourish in Belgium. As in Germany, Belgian beer-making was originally the province of monastic brewers. Regular political upheavals, however, meant that many of the old foundations eventually disappeared, so that today's Belgian abbey beers are largely brewed either in monasteries that have been re-founded since the travails of the sixteenth through eighteenth centuries, or are simply beers made "in the style" of monastic ales—which can cover a lot of waterfront. One special category of Belgian abbey beers comprises those with the Trappist designation, meaning that they are brewed in one of six monasteries belonging to the Trappist offshoot of the Cistercian Order, a splinter group that originated in France in the seventeenth century. Now eleven beers worldwide bear the coveted "Authentic Trappist Product" label, although all except one of the six Trappist brewing monasteries in Belgium today were established after 1835.

For a small country, Belgium boasts an amazing array of beers and styles. "Dubbel" and "tripel" were originally Trappist designations for heavy-bodied, fruity brown ales with ABVs of 6 to 8 percent and 8 to 10 percent, respectively, though tripels are typically more golden in color nowadays. Belgian amber ales are generally comparable to English pale ales, if often denser, maltier, and more alcoholic; blondes are in the same style range, though lighter in body and color (if not necessarily in ABV). "Champagne beers" undergo a second fermentation in the bottle; Flemish reds are inoculated with a *Lactobacillus* culture. Saison or "farmhouse-style" ales were traditionally lower-ABV beers made during the harvest season for thirsty workers on the farms of southern Wallonia. A longer-aged version, brewed originally for workers in the industrial north of Wallonia, is heavier in body and is known as *bière de garde*. But beware: casting tradition aside, many modern saisons have ABVs in the 5 to 8 percent range. One celebrated Belgian specialty: lambic beers that are brewed from wheat using wild yeasts, with extended aging. Fruit may be added—cherries to make kriek, raspberries to make framboise, and peaches to make pêche. If sugar is used to initiate fermentation the result is faro, while the slightly sparkling gueuze, the original sour beer, was traditionally made by blending younger, incompletely fermented batches with older ones, and allowing those wild yeasts to finish fermentation in the bottle.

Belgium, in short, is a wonderland for beer enthusiasts. The sheer variety of ales speaks to beer's deep historical roots in the area, although given the many disruptions the country has experienced over the past few centuries, it is hardly surprising that today's versions of most historic styles are not exact replicas of their ancient models. Not all are to everyone's taste; but all are interesting, and it is quite hard to find a poorly made Belgian ale. Yet the country also produces vast quantities of not very distinguished pilsner-style lagers. Oddly to us outsiders, most modern Belgian beer drinkers now opt for this lighter style: vastly more lager than ale is nowadays both produced and consumed in their country.

One reason that Belgium has such an endless variety of ales is almost certainly that it lies too far north to grow wine grapes. The same is true (or was until recently) of Britain, the other great ale-producing nation. Britain has a long tradition of making top-fermented beers going all the way back to Skara Brae, and the habit of brewing and drinking lager did not significantly penetrate the United Kingdom until the later twentieth century.

During the early Middle Ages Great Britain was awash in small beer, a valuable source of nutrition (and safe hydration) that was often made by reusing the malt from previous brews. Initially, the women known as alewives seem to have run the ale houses that sold beer made on the premises, but before long, men began muscling in on the business. By the fourteenth century, male brewers were forming guilds, and though they mainly produced for their own establishments, they also supplied others—foreshadowing the later system of tied houses. This development apparently necessitated an early form of consumer protection, in the form of "ale conners" employed by municipalities to assess the strength of the product sold as well as to establish fair pricing for tax purposes. Individuals sometimes had to be strong-armed into taking this post, maybe because the beers the conners were obliged to sample were not always of good quality. Sadly, the story that conners had to sit in a puddle of beer to determine if it had enough extract to stick their leather pants to the bench is probably apocryphal. But spoilage would have been a particular issue in medieval Britain because preservative hops took a long time to catch on there: they didn't become a regular component of British beers until the sixteenth century.

By the turn of the eighteenth century the larger British brewers were turning out a new style of ale, known as porter (allegedly because of its popularity among market porters) that was highly hopped and made from darkly roasted malts. With a typically high ABV of 6 percent or more, and made with the aid of early scientific instrumentation such as thermometers and hydrometers, this was the first beer that could be produced and distributed as an industrial product. The economies of scale enjoyed by the large brewers who manufactured it soon made it impractical for individual hostelries to brew their own beer.

Kilns for roasting barley were traditionally fired by wood or coal, producing a rather dense, smoky malt. Porters were correspondingly heavy, dark brews. But rapid technological advances in the early eighteenth century made clean-burning coke much cheaper and more widely available. This development paved the way for the large-scale production of the lighter-colored malts that were the basis for a burgeoning new category of pale ales—and, as we have seen, for Josef Groll's pilsner lager.

One highly significant variation on the pale ale theme was the India pale ale (IPA) that was produced specifically for the fledgling British Empire. The hot Indian climate made it impractical to brew beer on the spot, yet the sweltering British merchants and sundry adventurers wanted more than the pungent and often downright dangerous arak palm wine that was all Indian tradition had to offer. The market they offered was potentially hugely lucrative, but transporting English ales to India involved a long and arduous sea journey that traditional beers rarely survived in ideal condition. The solution to this problem was to increase the alcohol slightly, and the hopping greatly, based on an existing style known as October beer.

Octobers were strong pale ales of a variety greatly beloved by the landed gentry (as A. E. Housman put it, "There's many a peer of England brews / A livelier liquor than the Muse"). Typically, these beers were aged in the cellars of grand houses for two years, but the slightly briefer voyage to India proved to accomplish the same feat—and more. The ale arrived in the tropics not only bright, fruity, and refreshing, but often slightly sparkling, due most likely to a secondary fermentation that occurred in the barrels via the activity of *Brettanomyces* yeasts (see Chapter 8). Vast quantities of IPA were exported to India, and to Australia beyond; and by early in the nineteenth century a less alcoholic version was already being marketed to domestic consumers. Lighter

IPAs of this kind were also exported to the Continent, where they found a receptive audience: Bass ale appears alongside bottles of Champagne in Édouard Manet's great 1882 painting *A Bar at the Folies Bergère*.

In Ireland, porter developed in its own direction. When Arthur Guinness started his Dublin brewery in 1759, the state of Irish beer was reportedly rather dire. Guinness's response was to up the game, and by the end of the century he was concentrating on the production of a reputedly excellent porter that soon took over the market. Twenty years later his successors were brewing a very dark "superior porter" that eventually evolved into the "extra stout" version that became internationally famous, with its almost black color and slightly burned flavor. Guinness was helped along in its dominance of the dark ale market when the British authorities banned the heavy roasting of malts as an energy-saving measure during World War I. Production of porters and stouts in England plunged, leaving the field open to the Irish. And the effects of a taxation system based on ABV lingered: the weaker and significantly cheaper ales marketed as milds and bitters prevailed in the British market through the later nineteenth century, and well into the twentieth.

Just before World War II, Watney's Red Barrel went on sale in England. It was the first stabilized and artificially carbonated ale, and it was served under pressure from aluminum kegs. Other brewers followed suit, and the ubiquitous beer engines that had been used to raise living beer from the casks reposing in pub cellars began to disappear from bars across the nation, to be replaced by puny taps. The new ales were easier both to transport and to serve than their predecessors had been, but many traditional beer drinkers were dismayed by their lack of character. We will return to the consequences of this later; but meanwhile, in the United States the effect of Prohibition on the beer industry was more severe than anything wrought in England by taxation or world war.

America might be a puritanical nation, but it was founded on beer. Sailing along the Massachusetts coast in 1620, the Pilgrim Fathers decided to come ashore before reaching their planned destination of Virginia specifically because the *Mayflower* had run out of ale. Not long thereafter John Winthrop, newly appointed first governor of Massachusetts, was dispatched to the fledgling colony on a ship crammed to the gunwales with ten thousand gallons of beer, leaving barely any room for him. In due course, the signing of the Declaration of Independence was celebrated with beer in copious quantities, in the very Philadelphia tavern in which Thomas Jefferson had drafted the document.

All this beer was ale, of course; but in the middle of the nineteenth century a significant number of German lager brewers arrived, and American tastes began to change. The Germans discovered ideal brewing conditions in the northern Midwest, including abundant ice from the Great Lakes that eased the lagering process. Well before the end of the century Milwaukee alone was brewing half the nation's beer, mostly in the pilsner style, though a few breweries elsewhere bravely clung to ales. Because the local barley was somewhat different from the European strains, some brewers also began to experiment with using rice or corn in the mash to mimic more familiar flavors.

Then the blow fell. With the advent of Prohibition at the beginning of 1920 legal beer production ceased in the United States, and the nation's drinkers switched to more easily smuggled spirits. When the ban was finally lifted in 1933, the brewers were ready; but their supply chains had been disrupted, demand exceeded reasonable supply, and a lot of inferior product came onto the market. Consumption consequently sagged, and many brewers went under or merged, resulting in an industry increasingly dominated by giant brewing concerns— a trend that inexorably continues today. In search of economies and profits, the large brewers turned to barley alternatives and increasingly relied on advertising to sell their product. The growing availability of refrigeration helped; if your beer was ice-cold, nuances of flavor became less important.

By the middle of the twentieth century, mass-market American beer had become a pretty tedious product, and even popular imports such as Heineken were cast in the same mold. After a while, some British bitter keg ales gained a cult following in specialist bars in large American cities, but the American audience proved somewhat resistant to drinking "bitter" beers. Eventually someone had the bright idea of reviving the IPA name for ales that, truth be told, were ghostly imitations of the lavishly hopped originals.

The legacy of Prohibition was thus the dominance of industrial beers that were drunk ice-cold. That was bound to provoke a reaction, which duly came in the form of the craft beer movement that began to emerge in the 1970s. With no heavy hand of tradition to inhibit them, a younger generation of American small-scale brewers became, within a couple of decades, the world's most imaginative practitioners of their ancient profession. Under the looming shadow of the international giants, they have created what the English beer writer Pete Brown has memorably described as an American Beervana.

4

Beer-Drinking Cultures

In 1967, South Australia became the nation's last state to abandon the 6:00 p.m. "last call" at pubs, which had given rise to the infamous "six o'clock swill." Intrigued to know more about the beer that homeward-bound Adelaide workers had imbibed daily during their legally permitted seventy-five minutes (between the end of the workday and their wobbly exit from the pubs at 6:15, the official closing time), we acquired a bottle of the "Original" pale ale from South Australia's largest and oldest brewery, which made its first lager only in 1968. After rolling the bottle to awaken the live yeast within, we encountered a slightly murky ale that was light amber in color to begin with, but darkened a bit as more sediment settled out. The modest head dissipated quickly, and on the palate the beer was unassertive and only slightly hoppy. Still, not a bad drop for a quick swill on a hot Australian afternoon.

It's hardly surprising that the customs and rituals surrounding the consumption of beer are not only as diverse as culture itself, but also uniquely reflective of place. Indeed, you'll never fully understand any beer-drinking society without understanding its relationship to the beverage—conflicted as that relationship may be.

Back in the 1960s one of us stayed with friends in Saint Paul, Minnesota. Every Saturday, all the men in the extended family—and every family in the neighborhood, for that matter—went fishing in one of the numerous lakes nearby. Whether you felt like it or not, you would get up before dawn, pile into the truck with all your gear, and drive to Hamm's Brewery. Staff were on hand there to sell you as many cases of their product as you could carry. Then on to the fishing. As the sun rose, you scrambled onto a pontoon boat and pushed off into the middle of a flooded gravel pit, threw your lines overboard, and lowered your cases of canned Hamm's into the water to keep them quasi-cool. After shivering at first, you broiled aboard your shadeless craft through the middle of the day, vainly fighting dehydration with beer after beer—each warmer than the last—until at last the sun mercifully began to set, and you and your meager catch could honorably head back to the shore and the womenfolk.

To an outsider, this hardly seemed to be the optimal way to spend a precious day off, and even for a dedicated fisherman, a few tiddlers hardly seemed much of a reward. But naturally, none of this mattered. For above everything else, the jaunt was a social ritual through which men bonded and maintained their friendships. And it was all made possible by the beer. Certainly, as uninspired as Hamm's product might have been—especially as it warmed up—it was a lot easier to bond over than a peculiarly tedious form of fishing.

In 1963 Saint Paul, beer—however bland—was essential social cement. In contrast, bars were mainly for loners and losers. In those days the average American bar had yet to recover from the excesses of the immediate pre-Prohibition period during which the number of drinking establishments had tripled in a mere quarter-century—even as, tellingly, beer had steadily lost ground to hard liquor. And once drinking establishments had become legal again, the proliferation of supermarkets and the ready availability of home refrigeration had conspired to keep away the bars' former beer-drinking customers. Encouraged by expensive marketing campaigns, those good citizens began consuming endless canned and bottled beers at home—and out on lakes—with family and friends. Accordingly, the bar as an institution became ever more peripheralized, and an entire drinking culture perished. The German cultural historian Wolfgang Schivelbusch put it nicely, noting that the camaraderie of the bar—the toasting, the joking, the conversation, the standing of rounds—crumbled, as taverns ceased to be the preferred middleman between brewer and consumer. The typical bar became a dark, dank, sticky-floored place where stools were mostly occupied by those fleeing domestic strife, or who had no other place to go. Several decades passed before such dives became a relative rarity, and the American bar scene began to thrive again.

Such marginalization of the bar never happened in Australia, a hot country in which the advent of refrigeration had made the coldness of (mainly lager-style) beer an even more highly prized attribute than it is in the United States. That's why, as a rule, the hotter the region, the smaller the glasses in which Australian bars typically serve their beer—out of iced taps, of course. Giving the beer time to warm up just won't do. It's no accident that Australia is the home of the stubby cooler, that ingenious foam-rubber jacket designed to preserve your beer can's iciness—even if it contains a beautifully complex craft ale.

Beer is iconic all over Australia, and although a well-established wine industry has recently been making inroads, it remains the country's preferred beverage. Drinking is a seriously social matter, oriented much more toward gathering places like bars than it is in the United States. As the English observer Harold Finch-Hatton recorded back in the late nineteenth century: "In every class, it is not considered good form for a man to drink by himself . . . when a man feels inclined to drink, he immediately looks out for someone to drink with." What's more, should an Australian meet another "whom he has not seen for say twelve hours, etiquette requires that he shall incontinently invite him to come and drink." Not too much seems to have changed in the century and more since he wrote. But although the upshot is that in Australia a lot of drinking gets done, relatively low rates of drunkenness seem to result because the drink of choice is so often beer, and the purpose of drinking it is to enhance conviviality and conversation.

This highly social nature of Australian bar drinking is expressed most dramatically in the tradition of shouting, whereby everyone present is expected to buy a round. However reluctant you might have been to accept your neighbor's offer of a drink, once you've acquiesced you are on the hook yourself, and it's unthinkable to bow out until everyone in the group has shouted. Apparently, shouting acquired its name in Gold Rush times not because the shouter had to yell above the hubbub to order the next round, but because a successful prospector was expected to go out into the street and shout for his colleagues to join him in celebrating his good fortune.

Australia also provides an excellent example of how attempts to regulate alcohol consumption so often prove entirely self-defeating. Right around the time when the Temperance movement was gathering steam in the United States, before Prohibition struck, the various Australian states introduced the 6 p.m. closing of bars, which had previously been open until around 11 p.m. The underlying motive was invariably a narrowly moral one, aimed at reducing general drunkenness and rowdiness; but in most places the curtailing of drinking hours was publicly justified as an austerity measure associated with World War I. Early closing began in most states in 1916 or 1917, although Queensland held out until 1923, when it introduced an 8 p.m. closing. Tasmania returned to its senses in 1937, but more reasonable opening hours did not return to most mainland states until the 1960s.

The effect of restricting bar opening hours in this way was entirely predictable: the six o'clock swill. Workers streamed out of factories and offices at 5 p.m. and headed straight for the nearest bar. Once there, groups would order and swallow as many shouts as they could before last orders were called at 6:00, when they would line up and gulp down several more beers before drinking-up time ended at 6:15. Bars were enlarged to accommodate the crush by converting such civilized but space-hungry amenities as billiard lounges and darts platforms to additional standing room, and the heaving crowds of punters became rapidly inebriated as they drank far faster than they could metabolize even the relatively low-alcohol beer. Old-time barkeeps tell hair-raising stories of trying to keep up with demand, aiming their beer guns with one hand and operating the till with the other, then switching to the role of bouncer when drinking-up time was over. Suddenly the bars would be empty, as the streets filled with inebriated breadwinners staggering away toward the station and an evening likely spent passed out on the living-room couch. This was surely not the orderly kind of society that the wowsers of the Women's Christian Temperance Union had had in mind.

Nowadays you can get a drink in Australia pretty much whenever you want, and consuming alcohol has reverted to a much more civilized affair. Even the pub crawl, the evening-long bar-to-bar trek long beloved of heavier consumers, and the urban descendant of the "drink and bust" spree-drinking of thirsty colonial workers coming in from the bush, is said to be a fading tradition—though, like everywhere else, Australia is not without its alcohol-related problems. But overall, once reasonable drinking hours returned, ingrained tradition quickly trumped the unintended consequences of poorly thought-out legislation. Except in one area.

Crusading moralists never succeeded in getting alcohol completely banned in Australia, but they often succeeded in getting it banned for aborigines. This forced those dispossessed and basically unfranchised citizens to buy alcohol illegally, and in forms—mainly cheap and nasty liquor—that were easy to hide and carry. What's more, it forced them to consume this stuff away from the public sphere—the pubs and bars—in which social tradition would have obliged a measure of civility in consuming it. Social woes and personal tragedies inevitably followed.

After a disgraceful early colonial period during which European settlers discovered that a convenient way of dealing with indigenous Australians, unused to alcohol, was to ensure they were drunk as much of the time as possible, this opposite but equally destructive policy sadly remains alive and well. As recently as 2013 Australia's high court ruled that Queensland laws restricting the possession of alcohol by members of indigenous groups did not breach racial discrimination laws, on the breathtakingly paternalistic grounds that the measure had been taken "for the sole purpose of securing the adequate advancement of a racial group requiring such protection . . . [for its] . . . equal enjoyment . . . of human rights." Apparently, the court was not concerned that its ruling attacked a symptom of injustice, not a cause of it. And the justices certainly exhibited no awareness that, if society had really wanted to ameliorate the expressions of deprivation in many indigenous communities, there would have been many better ways of going about it. One of them would, of course, have been to bring drinking out into the open, and to allow it to perform for the original Australians the same role in enhancing social relationships that it has traditionally played for humankind the world over.

Although sake is of more distinguished lineage, lager-style beers are the alcoholic beverages most widely consumed in modern Japan. A lot of those lagers are "dry," made using a technology developed by Asahi Breweries in the 1980s to break down complex sugars in the wort. This manipulation makes those sugars available for conversion into alcohol (see Chapter 10), and the upshot is beers that are stronger and, to our minds, less flavorful than their western equivalents. As you might expect in a country as culturally self-aware as Japan, the consumption of beer there is often strongly ritualized. And in its own way drinking beer is often as intensely social as it is in Australia, even with what one might see as a local version of the six o'clock swill. After work, it was long customary for groups of white-collar salarymen to adjourn to drinking houses before returning home to their other lives, and the tradition was to start with a beer (or two, or three) to get the conversation going. Later in the evening the revelers might switch to other forms of alcohol, but whatever drink was responsible for the loosening-up, inhibitions were shed, releasing some of the strain that had inevitably built up while working within the stiflingly formalized hierarchies of Japanese business. Often everyone would get wildly drunk before the boss decided (sometimes only after the trains had stopped running) that it was time to go home and everybody could leave.

We wrote the last few sentences in the past tense because, even in Japan, things are beginning to change as old business norms break down and social media encourage participation in social networks other than business and family. This change has not just been in consumption patterns, for an energetic Japanese craft beer scene has emerged since the law governing brewing was changed in 1994. Nonetheless custom lingers, and in large cities you can still see red-faced salarymen staggering to the station late at night. Near those stations, sometimes under the arches supporting the elevated rail lines, you will find numerous *izakaya*, beer bars ranging from tiny to elaborate, that cater to individuals and groups who will typically linger more briefly before taking the train. Such convivial spots find their equivalents all over eastern Asia, perhaps most atmospherically in the hugely democratic and frequently open-air *bia-hoi* establishments that dot towns and cities all over Vietnam, with rickety chairs gathered around a dented aluminum keg sitting in an ice bucket. It is in such places that the assimilated lager beer truly comes into its own as a social lubricant, among people who have joyfully embraced the beverage.

Moving far to the west, the countries of southern Europe are usually viewed as wine-drinking societies; but in fact, a remarkably large amount of beer is consumed in those warm, southerly climes—hardly surprising, given the beverage's legendary thirst-quenching qualities. In Spain, for example, there is a beer tap in virtually every café, and the average Spaniard quaffs almost fifty liters of the cold stuff (usually lager of a modest 4 to 5 percent ABV) every year. Italians are not far behind. In countries where children are routinely introduced to (diluted) alcohol at a tender age, and ethanol is not a wickedly forbidden fruit during one's most formative years, the drinking of beer is simply a background fact of life. Consequently binge-drinking is rare, and the steady consumption of beer in moderation does what it does everywhere, facilitating social relationships and adding to a general sense of camaraderie.

Of course, beer can function similarly in places where it is most definitely regarded as special. And given that Germany is the supremely beer-loving society, and has been so pivotal in the beverage's evolution, it is hardly surprising that the world's most over-the-top expression of beer-drinking conviviality is hosted by the Bavarian city of Munich. We refer, of course, to Oktoberfest.

Oddly, the origins of Oktoberfest had nothing to do with beer. In October 1810, the soon-to-be King Ludwig of Bavaria celebrated his marriage to the Princess Therese of Saxe-Hildburghausen by having a new racetrack built, at a spot that duly became known as Theresienwiese (Therese's meadow), now usually shortened to Wiesn. In those days the Wiesn was just outside the city limits, though nowadays it is close to downtown. The wedding and the first race were fêted with a huge party, preceded by an elaborate parade in honor of the newlyweds. The entire extravagant occasion was such a success that it has been held almost every year since, wars and epidemics permitting.

After coming under the aegis of the city fathers of Munich in 1819 (a year after beer was first served), the Oktoberfest enterprise rapidly blossomed, even surviving Ludwig's abdication in 1848. Not coincidentally, Ludwig stepped down just four years after the infamous Beer Riots that fatally weakened his rule. Those riots had followed his imposition of a tax on the beverage (though his reign was more proximally ended by his scandalous involvement with the Anglo-Irish adventuress Lola Montez). The Oktoberfest subsequently became longer—it now spans a (literally) staggering sixteen days—and soon came to include all sorts of carnival-style attractions, including beer stands and an agricultural show. No longer linked to a royal marriage, the date of the celebration was also brought forward, to the balmier month of September.

Perhaps inevitably, this being Munich, eating and beer-drinking came to dominate the Oktoberfest, and various beer-related traditions began to accumulate: the ceremonial tapping and consumption of the first barrel, the brewers' procession, the costume parade, the fairground atmosphere, the beer tents, the heavy and almost unbreakable one-liter glass mugs, and many others. By the end of the nineteenth century the event was much as we know it today, with an extensive area of splendidly nostalgic fairground attractions concentrated on the eastern side of the Wiesn, while the enormous beer tents (nowadays not tents at all, but huge and gaudily decorated faux chalets in wood, boasting at best a symbolic canvas roof) are lined up along a broad avenue on the park's west side, fronted by a mind-boggling array of shacks selling memorabilia, candy, and fast food. Every one of the dozen largest tents, or *Festzelt*, can accommodate several thousand revelers in each of two daily sessions; the largest ever, in 1913, seated a mind-boggling twelve thousand. Each tent is run by a different brewery, with its own theme, traditions, and devotees. And although Oktoberfest is by now an international institution, annually attracting some seven million visitors from all over the world, locals will proudly tell you that 60 percent of the clientele is Bavarian—a feat made possible by the dedication of a large group of fans who attend at least every other evening for the duration of the event.

The demographic in the tent we attended consisted almost entirely of locals from Munich, and it skewed strongly toward those who were both young and in traditional Bavarian attire. Doors opened at 4:00 p.m., and by 4:15 the mainly brass band, just audible above the general din, had launched into a loud and eclectic playlist in which traditional Bavarian numbers, surprisingly, were a minority. By 4:30 all the tables were full, and an overflow standing crowd was already cramming the periphery of the tent. Dirndl-wearing and lederhosen-clad servers were weaving among the tables, brandishing impossibly large numbers of huge foaming beer mugs—we spotted one clasping an amazing nine of them. The beer within the glasses (whichever tent you're in, it will conform to the *Reinheitsgebot*) is typically a coppery lager in the hoppy and highly malted märzenbier style that was traditionally brewed in March, with an eye to consumption in the late summer; and for Oktoberfest, the ABV is usually amped up to the 6 percent range. Ours was.

By 5:30 vast quantities of food—mainly the traditional roasted half-chickens—had already been served and demolished, and the serious business of the evening was beginning. By six o'clock only a minority of revelers at the tables were still seated; most were already standing on the benches on which they had been sitting, or—although we were told this was bad form—on the tables themselves, now vacated of plates if not glasses. The noise and energy levels increased as the crowd sang along with the band, often linking elbows and swaying dangerously to the music. As time passed, people sang and conversed more loudly, ordering and swigging more beer. And while some roisterers eventually began to stumble, disaster was discreetly avoided as skillful handlers removed the very few who became incapacitated. Well before closing time at ten you were not only hoarse and slightly light-headed, but best friends with the multitude of strangers around you, even as the more dedicated carousers were already girding their loins for the rigors of the next day.

So what is Oktoberfest all about, beyond being the supreme cultural expression of Bavarian identity? As with most traditions, nobody seems any longer to care that the origins of the festival lay in a presumably rather stuffy (and beer-free) nineteenth-century wedding reception. Those who participated in the earliest beery Oktoberfests doubtless saw winter just over the horizon, and were happy to grab any opportunity for a blowout before wintry austerity set in. But today, when life is no longer ruled as it was by the seasons, Oktoberfest still offers an opportunity to combine two Bavarian obsessions: beer and *Gemütlichkeit*, a peculiarly German concept denoting a state of being that lies pretty much equidistant between good feeling and good fellowship. It is about how you yourself feel, but it is also about being in company. It describes the kind of well-being that you can experience only with others around; and, to judge by Oktoberfest, the more the merrier. Which, of course, brings us back to the social role of beer as a unique facilitator of bonding among friends, and as a breaker-down of social barriers among strangers. At the crowded communal tables of the Oktoberfest beer tents it's unthinkable that you won't start talking to your neighbors, or to the people at the table behind you. Or two tables away.

Although it is now celebrated at venues around the world, Oktoberfest remains a uniquely Bavarian institution. You simply couldn't imagine such a tradition evolving in England, where—unless they belong to the fortunately diminishing breed of binge-drinker—people are instinctively much more reserved and indirect than you need to be to enjoy Oktoberfest to the full. Instead, to understand beer culture in England you need to visit a pub, or preferably many. Pub, of course, is short for public house, and that's exactly how the institution began. Medieval alewives brewed and served beer to the public in their own houses, and many a pub started its life as the reception room of someone's private home.

A successful alewife attracted local customers by brewing a better product than her peers, and inevitably her premises would become a place where people would gather not just to drink, but also to exchange gossip and conduct community business. In these places, drinking beer and socializing (and possibly conspiring) became pretty much synonymous (shades of ancient Babylon). Usually more upscale than those alehouses were the inns that provided food, drink, and shelter to travelers and other transients. From modest origins as small-scale enterprises, inns expanded in tandem with the economy and enjoyed a golden age in the eighteenth and early nineteenth centuries as roads improved and the long-distance coach trade grew. That age was ended by the advent of the railways in the second quarter of the nineteenth century, with the associated rise of the station hotel. From early times, inns were a focus of social activity: it was at the Tabard Inn in today's south London that the pilgrims in Chaucer's *Canterbury Tales*, as motley a cross-section of polite society as you'll ever encounter, assembled in the late fourteenth century for their pilgrimage to Canterbury. Either way, alehouses or inns, these centers of hospitality provided a hub for the community, bringing people—both friends and strangers—together. And both kinds of establishment were founded on the beer in which the country was awash, though inns sold wine and spirits as well. A third category of watering hole, the tavern, had been around since Roman times, but, in keeping with the preferences of their long-departed founders, taverns specialized in wine.

Gradually, the distinctions between these various kinds of hostelry began to erode, giving rise over time to the hybrid form we know today as the pub. Establishments in the innkeeping tradition typically offered rooms to travelers, whereas the descendants of alehouses did not. But all pubs continued to provide a focus for social gathering—although in rural areas they tended to cater to the community at large, while in the burgeoning cities they often became sorted by clientele. As those cities grew, demand for beer soared. It was met by large breweries that enjoyed economies of scale, and that were enabled by improving technology to ship their beer over ever-greater distances. This greater geographic footprint in turn required reliable means of distribution, not only along the expanding network of canals (and later, railways), but also at the point of sale. Accordingly, big brewers eventually began snapping up pub after pub, resulting in the system of tied houses familiar to those of us of a certain age. These were pubs directly owned by the brewing giants, or under contract to sell only their beer. Soon only a minority of pubs were free houses, at liberty to sell anyone's beer and offering a choice to the discriminating consumer.

The thirsty industrial workers who spurred the increasing demand for beer were almost exclusively male, and with women preferring to stay home, pubs began to lose their status as centers of community life. Perhaps worse, throughout the nineteenth and twentieth centuries they were battered by wildly capricious taxation and licensing laws, the upheavals of one war after another, the general disfavor of the upper classes, adversely changing demographics, attacks from temperance zealots, smoking bans, and various other malign influences. In the wake of World War I, demand for beer plunged dramatically; and by the time World War II came around pubs were generally in a rather sorry state, serving weak and often characterless ales to a typically downscale clientele.

World War II changed all that. As bombs rained down on England, pubs once again became the symbol of community spirit, providing places for people to seek mutual support and to relieve the stresses of wartime life. But the beer itself suffered, as shortages of raw materials caused its quality to decline. Perhaps even worse, the brewers' newfound preference for making and distributing keg ales in place of the harder-to-ship and more laborious-to-keep cask-conditioned variety assured that, even as the war began to fade in memory, English punters would continue to drink relatively weak and uninteresting beers. To cap it all off, in keeping with the general listlessness of the postwar period many pubs remained rather cheerless and dilapidated.

No wonder, then, that during the latter part of the twentieth century pubs' sense of purpose began to waver as other avenues of public amusement proliferated. Almost as bad, heavily advertised draft and bottled lagers began to oust indifferent keg ales as a generation of drinkers turned over. In response, many pubs began self-consciously positioning themselves for specific sectors of the market. Some opened their arms to the family trade. Some became gastropubs, barely distinguishable in some cases from a restaurant with a good beer list. Some installed giant screens for sports fans. Some, god help us, began billing themselves as theme pubs. And some, coming spectacularly full circle, became brewpubs. Still, despite this identity crisis, the pub hangs on as a core social institution that is necessarily as diverse as the society that supports it.

Pubs at their best offer welcoming environments (preferably laden with Victorian woodwork) in which everyone, regular and stranger alike, can feel at home. And with English beer typically in the 3 to 4 percent ABV range, they are places where, male or female, you can drink and converse steadily all evening long. At their worst pubs can be seedy, depressing places in which most sensible people can't wait to finish their pints and leave—but one positive consequence of the steady drumbeat of recent pub closures is that such places are becoming much fewer and farther between. Looking forward, it is likely that pubs will increasingly make a conscious attempt to attract customers, whether by offering a more interesting range of beers (a feature made much easier by the blossoming of the British craft beer movement), by serving better food, or—by any number of stratagems—providing a more attractive environment. Whatever improvements the pubs embrace, though, their patrons show every sign of adhering to time-honored patterns of behavior. The buying of rounds may be less formalized in England than it is in Australia, but it is still a very good idea to get your round in early.

For all its trials and tribulations, then, the pub endures. But it endures by adapting. For the same reason that every generation famously needs its own translation of Homer, every beer-drinking society needs its own kind of environment in which to gather and drink socially. British pubs of tomorrow will not necessarily all be like those we know today, but the institution—and its many equivalents in other countries—will surely be around for as long as there is beer on the planet. Which is to say, for as long as there are people around to brew and drink it.

PART TWO

Elements of (Almost) Every Brew

5

Essential Molecules

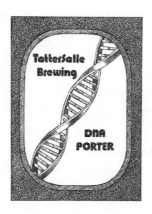

With no molecularly themed beers currently on the American market, this beer was another we had to make ourselves. We wanted an ale with a satisfyingly dense stew of molecules, so we settled for a strong porter. Into the brew went a complex blend of chocolate, crystal black, and wheat malts, followed by chips from old Bourbon barrels, Scottish ale yeast, and Golding and Chinook hops. The result was a dense, dark, and creamy ale with a durable head, a pleasing sweetness, and a hint of whisky on the aftertaste. It had all the molecular complexity we were looking for.

In the next four chapters, we will introduce the four major ingredients of beer: water, barley, yeast, and hops. From a natural historical perspective, these beer components are a pretty miscellaneous bunch, but there is one basic feature that they all share: molecules, those tiny structures made up of atoms. So, as a prelude to looking at those four magic ingredients in more detail, let's take a quick look at those tiny molecules. Doing this not only will help us to understand why beer tastes so good, and has such remarkable physiological effects, but also will give us a chance to unravel the evolutionary histories of all those key players in the brewing game. Some of what follows might appear a little technical, and you can skip it if you wish; but having read it may come in handy.

First and foremost, beer and its ingredients are made up of atoms, which are combined into molecules. Like most things on this planet, beer is carbon-based, meaning it has a lot of carbon atoms in it. Atoms come in many varieties, but animal bodies are selective in the ones they use, with most living animals incorporating only six. Carbon is the second most prevalent of them, behind oxygen. The mnemonic OCHNPS can be used to remember these big six atoms—oxygen (O), carbon (C), hydrogen (H), nitrogen (N), phosphorus (P), and sulfur (S)—in the order of their abundance in animals. Yeasts use the same big six, as well as a lot of chlorine. Plants use the big six, but four other atoms are also important in their structures: magnesium (Mg), silicon (Si), calcium (Ca), and potassium (K). As a nonliving and less complex entity, water has the fewest elements of any of our beer ingredients (just H and O), but it can carry a lot of other compounds around in solution or suspension, a quality that turns out to be very important for brewers.

All these atoms are incredibly small, and even when they are combined with other atoms into compounds and molecules, the resulting structures are tiny. Consider the water that makes up most of your glass of beer. A single water molecule is about 300 picometers long, or 0.0000000003 meters. One small standard beer-tasting glass is about 6 centimeters in diameter, or about 0.06 meters. This means it would take 200 million water molecules, placed end to end, to stretch across it.

Other important molecules include xanthohumol, a component of hops that gives beer its bitterness. Xanthohumol, a flavonoid, is bigger than the water molecule—its chemical formula is $C_{21}H_{22}O_5$—so that only about 8 million of them would be needed to span our tasting glass. Domestic beers vary in the amounts of xanthohumol they contain after hopping, but a concentration of 0.2 mg per liter is fairly typical. A 300 milliliter glass of such a beer would accordingly have about

0.06 milligrams of xanthohumol; and while this isn't much in terms of weight, it means that there are over 10^{22} (1 with 22 zeros after it) molecules of xanthohumol in the glass.

Numbers like this are meaningless in terms of our daily experience, but they do give us one way to comprehend the sheer scale of the chemical universe contained in any glass of beer. As we move along and learn more about the chemical reactions that happen in making beer, this sense of scale can help us understand just how complex and reactive a substance beer really is. We will also see how molecular archaeologists have been able to divine the components of ancient beers from molecular residues left behind in the pottery vessels that once contained them.

Most germane of all to our discussion of the natural history of the ingredients of beer is the large and beautiful molecule known as DNA (deoxyribonucleic acid). This is the molecule of life, and of heredity. DNA is a complex molecule made up of smaller molecules (called bases, or nucleotides), which in turn are made up of atoms of carbon, oxygen, hydrogen, phosphorus, and nitrogen. Each nucleotide has four major working parts (Figure 5.1). The center of a nucleotide is a sugar ring (not to be confused with the nitrogenous rings discussed next). On one end of the sugar ring (the 5prime end) reside three phosphates. The other (3prime) end has a hydroxyl (OH) group attached. Finally, coming off the side are those nitrogenous rings (the "base" part of the nucleotide), which give each nucleotide its identity.

There are four kinds of nitrogenous ring structures in DNA, and hence four kinds of bases. Two of these bases have two rings, and the other two have a single ring. The two-ringed nucleotides are called adenine and guanine, and the single-ringed nucleotides are called thymine and cytosine. They are known for short as A, G, T, and C, respectively. One of the three phosphates on the 5prime end likes to bond with a 3prime OH, so that the DNA strand has a direction to it: 5prime to 3prime. DNA is double-stranded, rather like a ladder with long parallel sides connected by rungs, and it also turns out that the side bases,

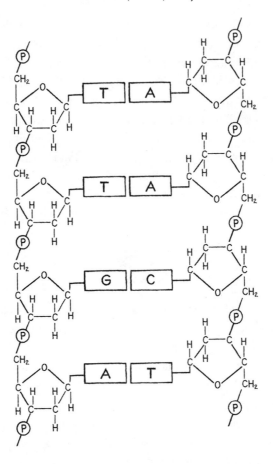

Figure 5.1. Diagram of a short stretch of double helical DNA. There are four bases or nucleotides on each of the two strands of the double helix. The base pairing of the nucleotides (A with T, and G with C) is also shown. The left strand runs top to bottom from 5prime to 3prime, and the right strand runs top to bottom from 3prime to 5prime.

those two- and one-ring additions to each strand, will stick to each other in specific ways. An A always likes to stick to a T, and a G to a C. This phenomenon is called base-pairing, and the bases are said to complement one another. This complementarity is critical for understanding both the beauty and the logic of DNA.

Imagine two complementary strands of DNA, twenty bases in length. They will stick tightly to one another, and coil around each other in a double helix; and it is the order of the bases along them that will determine each individual molecule's function. DNA works in pretty much the same way as our western alphabet, except that the "words" are only three letters long, and of course there are only four letters with which to work. Considering all possible combinations, sixty-four different words can be created in this way. The three-letter words code for amino acids, which are the molecules that make up the proteins which are building blocks of living things. Twenty amino acids are involved in making up proteins, so there are forty-four excess three-letter words in the DNA alphabet. Those extra three-letter words are used as redundant back-up codes for several of the amino acids, and three of them serve as "periods" at the end of a sequence of words coding for a protein. For instance, the amino acid proline (P) has four three-letter words that code for it — CCC, CCA, CCG, and CCT.

The sixty-four three-letter words form what is called the genetic code, and it is highly expressive. For example, each of the amino acids is different in terms of its size, charge, and attraction to water. The arrangement of the different charges, size, and hydrophobicity collude to add a unique three-dimensional, folded structure to the two-dimensional linear sequence of each protein specified by the DNA. It is this three-dimensional structure that, along with other molecular aspects of the sequence, usually dictates a protein's function in the organism.

The DNA sequences that code for proteins are known as genes. The barley genome, to take just one example, is five billion base pairs long, nearly double the size of our own human genomes. The 26,159 genes in the barley genome (we have a paltry 20,000) are arranged in linear strings of genes on seven pairs of chromosomes (we have twenty-three pairs). Chromosomes come in pairs in sexually reproducing organisms because every individual gets one member of each pair from its mother, and the other from its father. And what this does is to introduce variation into populations through recombination, the physical exchange of the DNA from one member of a chromosome pair to the other.

Figure 5.2 illustrates a specific region of a chromosome, for six individual chromosomes from two different populations. For most of the DNA sequence, the six chromosomes are the same. But there is

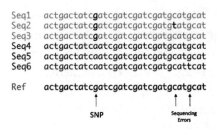

Figure 5.2. The SNP discovery. Sequences for six chromosomes are shown from two populations (light gray lettering is population 1; darker gray lettering is population 2). The reference sequence (Ref) is given below in black. The "SNP" arrow points to an SNP, while sequencing errors are indicated by the other two arrows.

one position in three of the chromosomes from the top population that is different from the three chromosomes from the bottom population. This variant spot is called a single nucleotide polymorphism (SNP), and it is the currency of modern genomics.

Sequencing a whole genome is not easy, not least because most whole genomes are billions of bases long and need to be sheared into little pieces for analysis. In fact, many published genomes that we consider whole have some gaps in them because of the random shearing of genomes into 100 billion to a trillion short sequences. The reason the pieces of DNA are sheared randomly is that you want some fragments to overlap with others, so that the whole can be put together from the overlapping bits, rather like creating a huge daisy chain of sequences (Figure 5.3). All of this is incredibly computationally intense, and sometimes the matches are incorrect. But if one already has a sequenced genome (for instance from one variety of barley), then ascertaining a new sequence becomes a lot easier, because that genome can serve as a scaffold on which to build the genomes of its close relatives. Sequencing a genome without such a scaffold (called a "reference sequence") is known as *de novo* sequencing. Fortunately, for barley, hops, and many

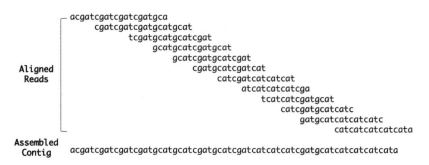

Figure 5.3. Twelve short read DNA fragments of up to twenty bases each, aligned so that they assemble a sixty-seven-base DNA contiguous sequence.

yeasts there are reference sequences for whole genomes that make this easier kind of sequencing possible.

The object of all this is to detect and categorize as many SNPs as possible. Some of the sequencing machines on the market can generate billions of short fragments of DNA called short reads, and the technology has advanced to a point where researchers can synthesize short fragments of DNA, or of proteins, and use these fragments as tools to help understand the genomes of organisms of interest. Since most of the bases in barley's five-billion-base genome will be identical from one plant to the next, researchers have developed targeted sequencing methods that focus on just the parts of the genome that have SNPs that differ (are polymorphic) over a wide range of individuals.

Once a computer has identified the sequences in which the important SNPs lie, short pieces of DNA are synthesized to match the sequences identified by the computer—but with a twist. Each SNP has five pieces of DNA synthesized, in which the SNP position contains a guanine, an adenine, a thymine, a cytosine—or nothing—to match the different possible scenarios for the actual barley DNA. The five short DNA pieces (each one detects a different base or its absence) are attached to a chip that is about the size of a quarter. The area that each piece takes up on the chip is so tiny that several hundred thousand fragments of DNA can be attached to a single slide, each in a specific location that is tracked by a computer. At this point, the DNA of the barley

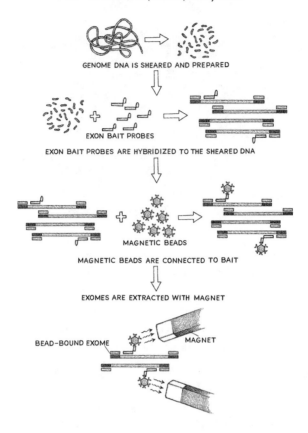

GENOME DNA IS SHEARED AND PREPARED

EXON BAIT PROBES

EXON BAIT PROBES ARE HYBRIDIZED TO THE SHEARED DNA

MAGNETIC BEADS

MAGNETIC BEADS ARE CONNECTED TO BAIT

EXOMES ARE EXTRACTED WITH MAGNET

BEAD-BOUND EXOME   MAGNET

Figure 5.4. Targeted sequence capture using beads. The small stippled lines are capture sequences with a bound biotin molecule (bent line). All other lines represent target sequences. The round objects with small protrusions are magnetic beads that bind to biotin. The magnet at the bottom extracts the captured sequences attached to the beads.

individual you want to sequence is cut up, labeled with a tiny fluorescent molecule, and allowed to react with the chip. Because of the DNA's desire for complementarity, all the pieces of barley DNA seek out the places on the chip that are 100 percent matched to them.

The chip is then washed and viewed under a special camera that can visualize the microscopic fluorescent spots on the slide. The camera will detect where the barley DNA hybridized, and hence will identify the base in that position. A variant approach uses the same protocol up to the chip step, but instead attaches a molecule called biotin to the short pieces of synthesized DNA where the SNPs reside. These small pieces of DNA are then used to "capture" the SNPs in the target genome. After thousands of probes complementary to each region of interest are placed on the chip, small magnetic beads bearing molecules that will bind biotin are mixed with the DNA. Any piece of double-stranded DNA bearing biotin will bind to a magnetic bead. A magnet is then used to separate all the beaded, biotin-containing molecules from the rest of the mixture. All these captured pieces of DNA without SNPs of interest are washed away, and the remaining DNA can then be sequenced using standard methods (Figure 5.4).

The targeted sequencing approach is perhaps the more accurate, because it allows for high resolution at about one-hundred-fold coverage (where coverage refers to the number of data points for a single SNP). Typically, several hundred thousand SNPs can be assayed by these methods. The panels for doing this are commercially available, and some are proprietary. Barley has several rapid sequencing arrays, with names like GeneChip®Barley Genome Array, the Affymetrix 22K Barley 1 GeneChip, and the Morex 60K Agilent microarray. Hops have not had a chip or an array developed yet, but the promise is huge, especially since a database (HopBase1.0) exists for the hops (*Humulus lupulus*) genome. Yeasts do have an array, called GeneChip Yeast Genome 2.0 Array, but since the yeast genome is so small, many researchers have simply taken to *de novo* sequencing the many strains of this organism. However the sequencing is accomplished, once you have those sequences they can be rapidly, efficiently, and cheaply exploited to uncover the genetic basis of the differences that brewers observe between the various strains and species of the same plant.

Perhaps the biggest challenge posed by genome sequencing is handling all the data. But it's a challenge worth rising to because those data, when properly interpreted, can yield a lot of information about the biology and natural history of any organism. If the relationships among species are the focus of a study, several different techniques can be applied to the sequence data. For instance, if we want to discover the closest relative of domesticated barley, or the closest relative of the ancestral hop plant, we can use methods that employ genomic data to generate what are called phylogenetic trees. We will go into greater detail about this in Chapters 6–9, but it's worth outlining the relevant methods here.

In determining how an agricultural organism was domesticated from a wild ancestral form, we can use the methods of phylogenetics, in which branching evolutionary trees are generated for groups of organisms based on how recently they had a common ancestry. With a phylogenetic tree in hand, one can make inferences about the relationships of one species to other species in the tree. If two species come from the same branching point on a tree, with no other species in between, they can be inferred to be each other's closest relatives, and can be called sister taxa. Another way to analyze genome-level data is to look at the dynamics of populations within species. Such population approaches will become particularly important when we look at the relationships among yeast strains, and barley and hops accessions. Using these approaches, one attempts to determine, first, whether the individuals under study are structured into populations. Once this is done, we can infer exactly how many populations we have. We can also use genome-level information to unravel how both natural and artificial selection have impacted the genome. These latter methods scan the entire genomes of individuals belonging to a given species to identify regions that reflect the footprint of selection. For domestic organisms like cultivated barley, this is tantamount to deciphering how breeders have gone after specific qualities in the barleys they have grown.

The methods that genomicists use are very visual. One of them is called principal components analysis (PCA). This statistical approach produces a number value for the difference between two strains, based on all the variables involved in the comparison. The two variables that explain most of the pattern of variation among the organisms analyzed are called principal components 1 and 2, and are represented on the X and Y axes of a two-dimensional graph. These values cluster the units of analysis in this two-dimensional space, so if we have four individuals, and we have the following distances for the first two principal components—A to B = 0.1, A to C = 0.5, A to D = 0.5, B to C = 0.5, B to D = 0.5, and C to D = 0.1—the PCA graph will show A close to B, C and D clustering with one another, and the two clusters lying far apart.

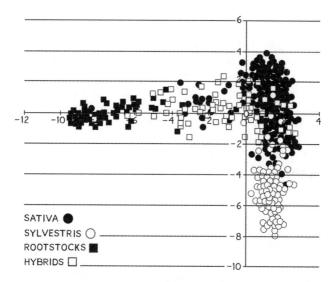

Figure 5.5. Principal component analysis of grape genomes. Solid circles are *sativa* strains (the subspecies of grape used in winemaking); rootstocks (used in grafting) are solid squares; subspecies *sylvestris* (the subspecies of grape that grows wild) are open circles; and hybrids are open squares.

As we will soon see, this approach can help give us a ballpark view of the overall relatedness of the individuals involved, and of the number of clusters in a study.

A study of the population structure of another plant important in alcoholic beverage production—grapes—gives us a good example of the approach, and of what we can expect from it. In this study, four groups of grapes were analyzed for 2,273 individual grape strains. The four groups were rootstocks, hybrids, and two subspecies of *Vitis vinifera*: *sativa*, which is the subspecies used to make most wines, and the wild-growing *sylvestris*. Note that there is some overlap of the four different clusters, and that those clusters are only made evident by the shading used in Figure 5.5. If we showed you the diagram with all the points in black, it is quite likely that you would not so readily infer four populations.

Accordingly, a more objective way of determining the number of clusters or populations present was developed by Jonathan Pritchard

Table 5.1

	Probability of inclusion in population A	Probability of inclusion in population B
Individual 1	100	0
Individual 2	78	22
Individual 3	22	78
Individual 4	0	100

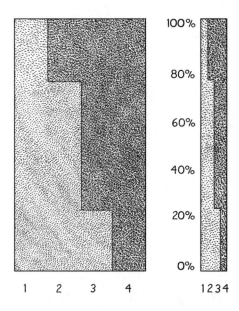

Figure 5.6. *Left:* Simple bar plots for four hypothetical individuals in a STRUCTURE analysis that have been assigned to two different populations (light gray is population A, and dark gray is population B). *Right:* the same bar plot compressed horizontally, which is how the structure diagrams are usually presented.

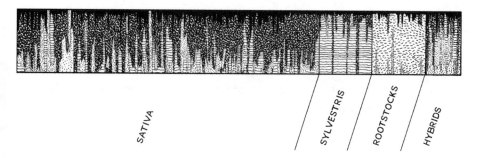

Figure 5.7. STRUCTURE analysis of 2273 strains of grape in the *Vitis vinifera* germplasm collection. The different shading represents the different kinds of strains in the analysis. Dark stippling represents *sativa*, long line stippling shows *sylvestris*, short line stippling represents rootstocks, and dotted stippling illustrates hybrids. This analysis is for K=6 (or six ancestral populations). Adapted from Emanuelli et al. (2013).

and colleagues. Called STRUCTURE, this approach is an iterative one, in which a model of population structure is simulated. The number of populations in the simulations is designated K. Using the genetic data, STRUCTURE runs the simulation for however many populations you think you might have. By comparing the statistics of each run for the different values of K, the approach can determine how many populations are most likely represented. With a good estimate of K, the approach can then assign every individual in a study to the K populations. Some individuals will be assigned with 100 percent probability to one of the K populations. But based on simple probability some individuals will be assigned to two populations, because of the effects of interbreeding across populations. Take an example with four individuals and K=2, with the assignments shown in Table 5.1.

Visually you will get the bar graph in Figure 5.6, which shows the percentages of assignment to the two populations.

Ramping this up to a large sample of grape varieties from different regions, the STRUCTURE approach gives us the population structure represented in Figure 5.7. Note that the assignment to population of origin for one of the subspecies of *Vitis vinifera* is difficult, or multishaded, both in subspecies of *sativa* (the subspecies of grape used in winemaking) and in hybrids. The assignments of rootstocks (roots of specific plants that are used in grafting) and subspecies *sylvestris* (the subspecies of grape that grows wild) are better delineated.

Observations like these are both intriguing and informative with respect to our understanding of the population structure of organisms, and as we will soon see, they are critically important for understanding the ancestries of domesticated barley, yeast, and hops.

6

Water

On the left, a bottle of our favorite pilsner from the eponymous town of Plzeň, renowned for the delicate softness of its water. On the right, a pilsner-inspired lager from Germany's Dortmund, with water about as hard as it gets in any classic brewing town outside Burton-upon-Trent. Granted that no two beers are ever identical, could we detect the influence of the water in the two brews? Well, the two beers differed like night and day. The one from Plzeň was golden, malty, and slightly sweet, while the Dortmund pilsner was steely, hoppy, and bitter. But in the end, it hardly seemed appropriate to ascribe those aggressive differences to the water. Dortmund is in northern Germany, while Plzeň lies close to Bavaria; and the two pilsners contrasted in exactly those features that tend to distinguish northern from southern German beers. The back label on the bottle from Dortmund declared, "ein pils bleibt ein pils" (a pils is always a pils). We wondered.

Now that the importance of yeast in brewing is common knowledge, the unsung hero of beer is the water that comprises up to 95 percent of the brew. If we take water for granted, we do so at our peril. As any proud resident of Plzeň or Burton-upon-Trent will assure you, the quality of the water from which it is made has a strong effect on any beer, even if it is often difficult to disentangle from other factors.

Water is an essential component of our everyday lives. In fact, our very bodies are about 75 percent water. Remarkably, we are talking about a very simple molecule here, H_2O. Just three atoms, some of which may be almost as old as the Big Bang at the beginning of the known universe. The Big Bang occurred about 13.5 billion years ago, with a flash of unimaginably intense heat. Within fractions of a second significant cooling began, allowing hydrogen (H) and helium (He) to form within about three minutes. The rest of the elements came later, although oxygen (O), the other elemental component of water, was recently shown to exist in a star system that lies a mind-boggling 13.1 billion light years away from oxygen-rich Earth, suggesting that the first oxygen atoms formed very long ago indeed. But water itself would not appear until much later.

According to the Dry Earth theory that was widely accepted until recently, water was initially brought to our planet by asteroid bombardment hundreds of millions of years after the formation of the solar system some 4.5 billion years ago, as the newly formed Earth "mopped up" the planetary debris lying in its orbit. But the study of an asteroid called Vesta that circles our Sun has been changing minds about this. Vesta has water, is of similar composition to Earth, and formed at the same time, implying that water formed concurrently on both bodies. In other words, water accreted on our planet the same time it did on Vesta, suggesting that our planet was wet from the beginning, and has stayed wet.

Still, this doesn't mean that the water we use in brewing our beer is 4.5 billion years old. Water is a very reactive molecule, and it is excellent at dissolving and combining with other chemicals and compounds. Because of this reactivity, the lifespan of an individual water molecule has been estimated to be about one thousand years, which is apparently about the longest a water molecule can go without reacting with some other compound and being torn apart.

Water may be a simple molecule, but it has physical behaviors that make it unique. Water has two hydrogen atoms and a single oxygen atom. Atoms themselves are made of smaller particles known as neutrons, which are not electrically charged; protons, which have a positive charge; and electrons, which are negatively charged. Mother nature is a pretty stern bookkeeper when it comes to these electrical charges, and molecules are more stable when the charges are balanced. Water is nicely balanced, and therefore chemically stable, because the two hydrogens in water both share electrons with the single oxygen atom, in what is called a covalent bond. But as part of their balancing act, water molecules have a positive and a negative end, making for some very interesting interactions with other molecules.

Water's reactive nature lies at the heart of its mysterious qualities for the brewer. Some metals react rather violently with water. Many readers might remember doing the exothermic reaction between metallic potassium and water in high school chemistry. This explosive experiment is always done outside, for very good reason. Even sodium will react quite violently with water, yielding the amazing sight of fire burning on a watery surface. On a milder scale, when they are immersed many molecules (among them sodium chloride [NaCl] and other salts) are pulled asunder into their constituent parts (Na^+ and Cl^-), which have charges. Those single molecules with a charge are called ions, and the water molecules that surround them tear them apart in the chemical process known as dissolution. Small molecules like sugars, and longer-chain molecules like complex carbohydrates, will also dissolve in water; but they are not pulled apart, and instead become distributed throughout the water matrix. This distinction is important to the brewer, because it means that water molecules will form very weak bonds with sugar without any structural alteration—something that is very important in making sugars available for the reactions with yeast enzymes that occur during fermentation. Figure 6.1 illustrates the difference between dissolution and dissolving.

With all this dissolution and dissolving going on, ground water inevitably carries around a lot of different kinds of ions that can help or

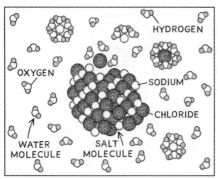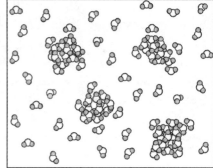

Figure 6.1. The difference between dissolution and dissolving. On the left, sodium chloride (NaCl), a salt, is dissoluted by tearing apart the NaCl molecule (large dark and small white balls clustered in the middle of the diagram) and surrounding the free sodium ion (Na) and chloride ion (Cl) with water molecules. In the diagram on the right, sugar is dissolving. The sugar molecules are not torn apart, but rather are surrounded by water molecules (there are four clumps of them).

hurt the beer-maker's goal of maintaining the optimal pH (alkalinity or acidity) of a fermenting beer mash. The pH is significant because the various enzymes that break down the components of grains like barley work best at specific acidities. Finally, water is important because its temperature can be regulated easily, and distributes evenly. No wonder biology texts often call water the solvent of life.

Water, by far the largest component in beer both as it enters and leaves the brewing process, also greatly influences any beer's taste. This is because it inevitably contains many more compounds than its simple basic molecules. Thus, even if it is only weakly acidic, any natural water source—a river, a lake, an aquifer—will pick up and incorporate dissolved calcium, magnesium, sodium, potassium, and other ions. In addition, because other chemicals—including, in the modern world, even hormones and antibiotics—leach into water supplies, most of the different waters you encounter will be quite heterogeneous solutions.

Table 6.1

Classification	parts per million (ppm)	milligrams per liter (mg/l)
Soft	less than 100	less than 17.1
Slightly hard	100–200	17.1 to 60
Moderately hard	200–300	60 to 120
Hard	300–400	120 to 180
Very hard	more than 400	more than 180

Even water that has been purified using activated carbon devices will have a lot of chemicals in it. Carbon filters are good at pulling out chlorine, phenol, hydrogen sulfide, and some other volatile and odoriferous compounds, and they can remove small amounts of metals such as iron, mercury, and chelated copper. But water purified with a carbon filter will still contain significant amounts of sodium, ammonia, and a host of other chemicals.

Water is often described as soft, very hard, or in between. The hardness of water is determined by its quantity of positive metal ions with a charge of +2. The most common of these in hard waters are magnesium and calcium. A very hard water will have a lot of positive metal ions in it, making it basic, or above 7 on the pH scale. The hardness of a given volume of water is usually measured in parts per million (ppm), while the concentration of the metal ions in water is quoted in milligrams per liter (mg/l). Like their names, hardness and softness are somewhat subjective; but the ppm and mg/l scales allow the classification shown in Table 6.1.

Tap water is notoriously variable from place to place, which makes where you do your brewing critically important. Each kind of brew will work better at a specific pH, and even in a single place not all municipal tap or industrial water supplies will provide water of even quality throughout the year. The source and quality of your brewing water are vital to the product you want to make, and many famous brewing cities have achieved their eminence because of the quality and consistency

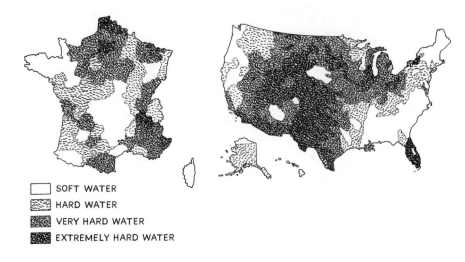

SOFT WATER
HARD WATER
VERY HARD WATER
EXTREMELY HARD WATER

Figure 6.2. Maps showing the location of major kinds of water hardness in France and the United States.

of the water available locally. Figure 6.2 presents examples, from two countries, of how water hardness/softness can vary from place to place.

Water's hardness is dictated by what the water has experienced since it fell as rain. Surface water from a lake or a pond is almost always quite soft. But groundwater that has traveled through rocks before it is pumped to the brewery will have picked up lots of minerals, especially if it has flowed a long distance underground. Water flowing through limestones, for example, will typically have acquired significant amounts of calcium and magnesium. Local geology will thus have a big impact in decisions about where to site a brewery.

The great brewing cities of the world exemplify this variation. Table 6.2 shows the water hardness for several European brewing towns.

Some of the cities with very hard water are centers for brewing heavy ales or stouts. It is no coincidence, for example, that Burton-upon-Trent, with its gypsum-rich water, became the classic center for brewing those durable IPAs. Brewers in other hard-water cities almost always nowadays treat their water to bring it into a softer range, something most easily done if the water is only "temporarily hard," with un-

Table 6.2

City	Water hardness (ppm)
Burton-upon-Trent	330
Dortmund	283
Dublin	122
Dusseldorf	104
Edinburgh	176
London	94
Munich	94
Pilsen	10
Vienna	260

wanted ions that can be removed by boiling the water in the presence of carbonate or bicarbonate, which will precipitate out the magnesium and calcium. On the other end of the scale, one city that sticks out for its exceptionally soft water is Plzeň, where the crisp pilsner lagers originated.

It is generally preferable to have a supply of softer water at hand when brewing, because it is more difficult to soften hard water than it is to simply add minerals to make a soft water harder. In general, if a hard water source is too much over 100 ppm, then it is good for brewing only a narrow range of different beers untreated. Four compounds are commonly added to make water harder, in order to emulate the ideal water for specific beers. Calcium sulfate (gypsum) or calcium chloride hardens water for lighter- to medium-bodied ales; calcium carbonate can increase hardness for dark beers; and magnesium is often added to mimic water thought to be ideal for a range of English-style ales. In most cases, then, less was more when it came to water hardness in brewing, and only relatively recently has water-treatment science allowed brewers to substantially escape from the limitations imposed by local geology.

Compounds on this planet can exist in three different states—gases, liquids, and solids—depending on their temperature and pressure. Water is the most versatile of them over a narrow range of temperatures, and it is one of very few that can be found naturally in all three phases. The solid phase forms when the liquid water molecules, which are packed very tightly and irregularly, and are bound together by hydrogen bonds, rearrange themselves into a regular lattice of molecules. Water is bizarre in being less dense in its solid than in its liquid phase: ice, as we all know, floats in water. This is because the lattice keeps the water molecules at specific distances from each other, making ice less dense than liquid water. Gaseous water, in contrast, is more typical in being less dense than its liquid form. When liquid water is heated, the relatively weak hydrogen bonds keeping the water molecules bound to each other are broken. As they break the water molecules begin to separate and to push away from each other, making the gaseous phase much less dense and forming steam.

While we are speaking of density, we should note that, at 22°C (72°F), liquid water weighs about 0.998 grams per cubic centimeter (8.33 pounds per gallon). This is water's specific density. When compounds such as sugars are dissolved in water, its density will increase. But what does density really mean here? In its simplest definition, the term refers to how much matter there is in any given volume of a substance. With respect to molecules, this definition translates to how tightly packed the molecules of a compound are. Very tightly packed molecules of a compound make a denser object than loosely packed ones do, which is why gaseous water is lighter (less dense) than either solid or liquid water.

A derivative of density, specific gravity, is critically important in brewing. Technically, specific gravity is the mass of the solution being measured, divided by the mass of an equal volume of water. This means that the specific gravity of any volume of water is, by definition, 1.0. Because specific gravity cannot be expressed in terms of grams, pounds, or milliliters, "points" are used to describe a liquid using this measure. The specific gravity point value of a volume of liquid is equal to the specific gravity minus 1.0, times 1000. So if the specific gravity of a volume of liquid is 1.0666, the point value of that volume is 1.066 minus 1.000, times 1000, or 66.6. The specific gravity of a volume of liquid will increase about 4 points with the addition of every 1 percent by volume of dissolved carbohydrates. Thus, if one adds carbohydrates until they compose 20 percent of the mixture's volume, the points will increase by about 80.

The carbohydrates/sugars in a beer will thus increase its specific gravity, an increase equivalent to the total amount of sugar available for the fermentation process. As a result, specific gravity can be used to compute alcohol content. Before fermentation (which we will discuss in detail in Chapter 10), the specific gravity of the brew (at this stage called wort) is called the original gravity (or OG). The specific gravity of the brew after fermentation has stopped is called the final gravity (FG). Water is such a wonderful solvent that it will absorb not only additives like carbohydrates and smaller sugars, but other compounds as well that also influence OG.

As the brew ferments and the sugars are converted to alcohol, the specific gravity of the brew rises because alcohol is less dense than sugar. The entire solution will therefore become less dense than it was before fermentation began. The FG will be less than the OG, and the drop in specific gravity will tell us something about how much alcohol was produced. Because specific gravity includes the impact of dissolved carbohydrates/sugars and other compounds in the brew, the before-and-after specific gravities don't tell us the alcohol conversion exactly, but they are generally quite accurate.

Most beers can be characterized by their OG to FG range, as shown in Figure 6.3. Some brews (heavy beers) start with fairly high OGs. These include doppelbocks, eisbocks, strong Scotch ales, Russian imperial stouts, Belgian dark strong ales, and barleywines, which have OGs in the range of 1.080 to 1.120, or 80 to 120 points. Only a few beers have OGs of less than 1.040 (40 points); these include standard ordi-

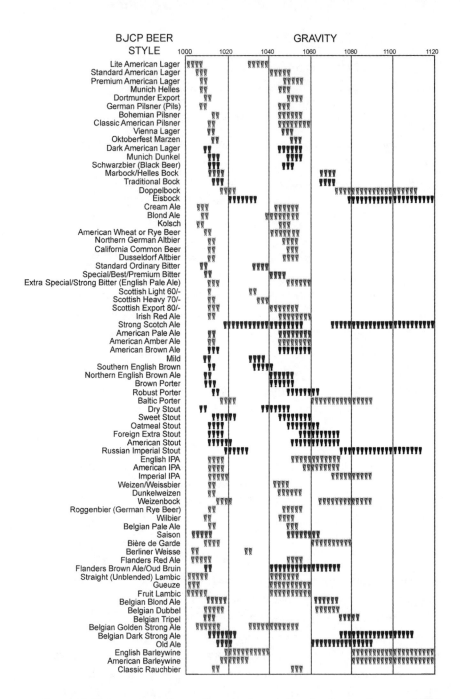

Figure 6.3. Various types of beers (from the Beer Judge Certification Program, BJCP) that illustrate the ranges for their original gravity (OG) and their final gravity (FG). There are two ranges as marked by the small beer glasses. The range on the left is FG and the range on the right is the OG. The different shades represent the overall darkness of the beer brewed for that type.

nary beer, lite American lager, Scottish light, Scottish heavy, Scottish mild, Scottish/English brown, and Berliner Weisse. Most beers start with OGs between 1.040 and 1.060 (40 to 60 points).

Water does some strange things with respect to the mass of objects placed into it. The Greek philosopher and mathematician Archimedes was supposedly one of the first to realize this. He was contracted by the tyrant Hiero to sleuth out whether his goldsmith was cheating by substituting silver for gold in the construction of one of his crowns. Archimedes is said to have realized how to figure this out when, sitting in his bathtub, he realized that his body displaced the water according to how much of it was immersed.

The journalist David Biello looked closely at this great story, and concluded that although Archimedes very likely did show that the goldsmith was cheating Hiero, most of the story's details—including that upon making his discovery the savant jumped out of his bath and ran down the street naked, shouting "Eureka!"—were probably not true. Nonetheless, Archimedes's principle of buoyancy is perfectly valid, and it has proven a boon to brewers. Employed to measure specific gravity before and after fermentation in both beer- and wine-making, it simply states that an object placed in water will experience an opposite force (buoyancy) that is equal to the weight of the water displaced by the object's submersion. Anything that floats, in other words, has a buoyancy corresponding to its weight. Picture this: we have liquid in a tall glass container. Let's say the liquid is water (with a density of one gram per cubic centimeter). Now let's take a cylinder that is one square centimeter in cross-sectional area and weighs ten grams. If we place this cylinder in the water-filled container it will displace ten grams of water and sink a corresponding amount. Because the cylinder is one square centimeter in cross-sectional area, it will sink exactly ten centimeters into the water (1 square centimeter times 10 centimeters equals 10 cubic centimeters) if it weighs the same as water. If the sinking cylinder is balanced to stand upright in the solution, we can easily measure how far it sinks by marking centimeter units on its side.

Archimedes confirmed Hiero's doubts about his cheating goldsmith by comparing the displacement of the fabricated crown to that of a chunk of gold weighing what the king wanted the crown to weigh.

Figure 6.4. A typical hydrometer.

He found that the displacement of the gold was greater than that of the crown, and so could infer that the crown was not pure gold. Changing the context a little, we can use Archimedes's principle to determine the change in displacement of the same object in two solutions of different density. So now imagine the cylinder described above in a solution that has a density of 0.95 grams per cubic centimeter. The cylinder still weighs 10 grams, and it will sink until it has displaced 10 grams of the new solution. But the displacement won't be 10 centimeters this time, because the density of the solution is now only 0.95 grams/cubic centimeter. Specifically, the cylinder will sink by 10 centimeters divided by 0.95 grams per cubic centimeter, or 10.53 cm. The cylinder will sink farther because the solution is less dense.

Now imagine that the first solution we want to measure is wort before fermentation, and the second solution is the same wort after fermentation. Before fermentation there is more sugar in the solution, making it denser than it will be after fermentation when the sugars have been converted to alcohol. All we now need to do is to make a cylinder appropriate to the job and we will have a tool for estimating the alcohol content of a solution after fermentation. This device is called a hydrometer, and it was not invented by Archimedes, but by one of his late-fourth-century successors, Synesius of Cyrene. A typical hydrometer is shown in Figure 6.4. The bulb at the end contains a set weight in grams, and it is the part of the device that sinks into the solution. The depth of sinking is given on the scale part of the hydrometer at the top, and it is read from where the surface (the meniscus) of the solution crosses the scale. We will look in detail in Chapter 10 at how measurements before and after fermentation are used to estimate alcohol content.

7

Barley

The three beers from the Abbaye de Notre-Dame de Saint-Rémy went into three identical glasses, from left to right the bottlings known as 6, 8, and 10. These respectively contained 7.5 percent, 9.2 percent, and 11.3 percent ABV, with the alcohol coming from increasingly generous additions of roasted barley malt and hard brown sugar. The colors ranged from a golden beige, through dark tan, to a deep nut brown, almost black. The tastes followed as night follows day: light and silky in the 6, mellow and gently sweet in the 8, and a thick, deep caramel in the 10. Oddly, we could hardly detect the large difference in alcohol content among the glasses. Not until we stood up, anyway.

Beer has been made quite possibly for as long as people have cultivated cereals—that is, for a very long time. Microscopic examination of twenty-three-thousand-year-old flint utensils found at the Ohalo II site in Israel has revealed the curious polish that usually forms only when the sharp stone blades, set into handles, are used to cut siliceous cereal stems. Remarkably, this was a dozen millennia before the last Ice Age ended, heralding the beginnings of settled life and the domestication of animals and plants in the Near East. This might mean that beer or beer-like beverages originated not long after humans figured out that grinding or pounding plant material yielded better tasting, sweeter food—a practice that goes back even farther in the archaeological record than Ohalo II. It's even been suggested that these beverages might have originated in an earlier era, since the chewing of grain (as is still done in making Andean chicha) adds salivary enzymes that convert starches into sugar, ready for fermentation. On this basis, it has been argued that beer might conceivably have been made in some form right back to the point at which our species began behaving in the modern manner, that is, around a hundred thousand years ago.

Many grains, among them rice, millet, corn, and sorghum, are used to make beers in different areas of the world, and all can be malted. But the key grain used in brewing the western-style beers that most of us drink is barley. This is not just a matter of historical coincidence: barley has what you might call an enzymatic toolbox that makes it the perfect brewing ingredient.

Like most grasses, barley has a fairly simple structure. This is shown in Figure 7.1, which portrays an entire plant from root to spike. For brewers, the spike is the important part of the barley plant, because it is where the barley seeds sit. The structure of the spike varies significantly among different strains of barley, and those different structures are keenly relevant to brewing beer. Spikes can vary in the number of rows of seeds they bear, in multiples of two: two, four, and six. And while more might intuitively seem better, six is not necessarily the preferred row number. Indeed, European brewers overwhelmingly prefer two-row barley. Significantly, six- and four-row barleys have a different enzymatic makeup from those with two, something we'll look at shortly.

The barley seed is layered, a property that is important for understanding why this is the preferred cereal for making beer. And it is the tiny sliver of seed tissue labeled the aleurone layer in the diagram that is critical in brewing. During the normal life cycle of a barley plant, the endosperm of the seed develops a large reserve of starch, destined

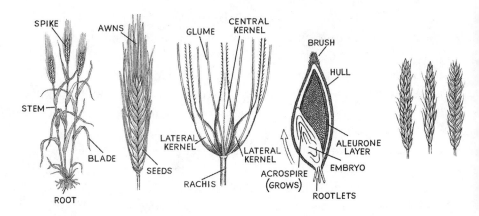

Figure 7.1. *Left to right:* The whole barley plant; a close-up of the spike; a detail of a barley kernel; a cross-section of a barley seed; and six-row, four-row, and two-row barley ears. The difference among the barley ears derives from the degree to which the spike is twisted, governing the number of kernels per row. The six-row barley has a two-thirds twist; the four-row has a half-twist, and the two-row barley is completely untwisted, so that all the kernels are symmetrical and straight, one row per side.

to power later development when the seed starts to germinate. In its original form this starch is not directly available to the seed for growth, but the aleurone layer contains a reserve of enzymes that are released when germination starts. Those enzymes promptly start to break down the endosperm boundary, exposing the starch granules inside the endosperm to other aleurone enzymes that break them down into sugars, primarily maltose (see Chapter 10).

Although other grains have an aleurone layer in their seeds, none has quite the capacity that barley does to break open the endosperm and turn starch into sugar. Accordingly, a brewer making beer primarily with rice or wheat will usually also add some barley. The process of getting the sugars out of the barley seed by starting germination is known as malting, and the maltsters who do it hijack nature's system by keeping the barley seeds from germinating until they want to make their malt. At that point, germination is artificially induced (Chapter 10).

The seed arrangements of six-row, four-row, and two-row barley varieties vary according to the degree to which the spike is twisted. This twisting governs the number of kernels per row. Two-row barley is completely untwisted, so that all the kernels are symmetrical and straight, one row per side. Six-row has a two-thirds twist to it, and four-row has a half-twist (Figure 7.1).

Most beer outside the United States is brewed from two-row barley, while New World brewers incline to the six-row forms. A question of taste may be involved here, because there are flavor differences between the two kinds. Barley can also be cultivated in both the spring and the winter, a major difference being that winter barleys require a process called vernalization (basically, cold exposure) to stimulate flowering during the late fall. If vernalization does not occur, winter plants will fail to produce a seed head. Most cultivated barley strains (known as landraces) fare better as spring crops than as winter crops, and right up until the 1960s most malting in Europe was done with two-row spring barley.

There are literally thousands of barley cultivars. A document entitled *Global Strategy for the Ex-Situ Conservation and Use of Barley Germ Plasm* summarizes both these and the many kinds of wild barley from around the planet. All are represented in collections of wild and cultivated landrace barley accessions established by the International Treaty on Plant Genetic Resources for Food and Agriculture (ITPGRFA), an agreement that governs their recording and arrangement in more than fifty institutions worldwide. The total number of accessions has now reached around 400,000. The biggest and most inclusive of these collections is in Saskatoon, Saskatchewan, at the Plant Gene Resources of Canada (PGRC).

Barley growers have kept good breeding records over the last century or two, so that the pedigrees of many of these cultivars are well known. Roland von Bothmer, Theo van Hintum, Helmut Knüpffer, and Katuhiro Sato have summarized them in their book *Diversity in Barley*. About 36,000 accessions exist from barley cultivars, of which 25,291 have pedigree information. The maps in Figure 7.2 show where these cultivars are grown, and where the wild strains in the collections (over

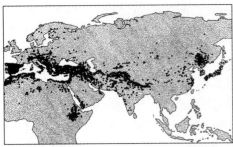

Figure 7.2. Maps showing the geographical origins of stored cultivars of barley (*left*) and stored wild accessions of barley (*right*). Each black dot represents an accession. Adapted from the Global Strategy for the Ex-Situ Conservation and Use of Barley Germ Plasm report.

12,000 accessions) come from. Because all Western Hemisphere cultivars derive from imports from Europe and Asia, this part of the world is not shown.

Not all accessions are used in brewing, and many are used exclusively in livestock feed production. But modern maltsters and brewers make use of many of them, and each year in the United States the American Malting Barley Association (AMBA) informs maltsters which strains are going to be the best for that year. In Europe, Euromalt serves as the clearinghouse for information about barley strains and malting, while in Australia, Malt Australia performs the same service. The recommendations of these associations differ from country to country. For instance, in 2017 Malt Australia accredited twenty-seven landraces, of which Bass, Baudin, Commander, Flinders, La Trobe, and Westminster were listed as the major players. Like Europe, Australia focuses mostly on two-row barley strains for malting and brewing. In the United States, the AMBA listed twenty-eight accredited landraces for 2017, including both two-row and six-row forms. Among six-row barleys, Tradition and Lacey appeared to be the most sought-after for 2017, while the two-row landraces most in demand, and accredited by AMBA, are ABI Voyager, AC Metcalfe, Hockett, and Moravian 69.

Rice, barley, corn, and wheat are all very similar in their basic anatomical structures. After all, they are all grasses, and quite closely related. Grasses are monocots, members of one of two major branches in the plant tree of life. During plant development, a region of the plant embryo called the cotyledon develops into the very first leaves of the plant. Monocots are the flowering plants that have only one such cotyledon region (members of the other great flowering plant lineage, the "dicots," have two). The monocots are very diverse, and together with grasses, they include lilies, palms, tulips, onions, agave, bananas, and several more major groups. Along with grasses, lemongrasses, sedges, and bromeliads, cereals like barley, rice, wheat, and oats belong to the division of the monocots called Poales.

Poales can be further divided into over forty groups that include maize, barley, rice, and lawn grass. These grasses are all members of the family Poaceae, and, within this family, barley is in the genus *Hordeum*. Depending on which expert you believe, the barley genus contains anywhere from ten to over thirty species. The name *Hordeum* derives from the Latin "horreo," for "bristle," referring to the pointy spike. The barley used to make most beer is from the species *H. vulgare*, another Latin name that means "common." The wheat and rice also often used in brewing are members of the family Poaceae as well, and have the genus and species names *Triticum aestivum* and *Oryza sativa*, respectively.

In 2015, Jonathan Brassac and Fred Blattner used genome-level DNA sequence data to look at how the thirty-odd species of barley are related to each other. It was clear that *H. vulgare* and two other species, *H. bulbosum* and *H. murinum,* formed a group quite distinct from the other thirty or so species in the genus *Hordeum*. This confirmed the traditional morphological grouping of these species together in their own subgenus. But doubt continues to hover over one entity that is classified in its own species by some taxonomists, and as a mere subspecies by others. This is (to call it by its subspecies name) *H. vulgare spontaneum,* a form considered to be the wild counterpart of all the cultivated landraces of *H. v. vulgare*. There is still no agreement on whether this wild barley—the closest thing we know to the common ancestor of the landraces—is its own independent species, or whether all the domesticated forms remain conspecific with it.

Because the landraces of *Hordeum vulgare* have gone through what plant breeders call domestication syndrome, we should expect that some of the traits in the domesticated strains will differ from their counterparts in the wild strains. And it turns out that in the landraces of barley the spikes are much less brittle than in the wild forms. The brittleness of wild barley spikes enhances the dissemination of the seeds under natural conditions, but for human barley growers the calculation is very different. You don't want the seeds to fall off when you harvest your barley; and ancient barley breeders seem to have indulged in a rudimentary form of genetic engineering by selecting plants that had a particularly strong spike structure holding the kernels together during harvesting.

The obvious question to ask now is, "Where did the barley landraces come from?" But before you can figure that out, you need to know whether barley was domesticated only once, or independently on several occasions, from multiple wild strains. Several studies have looked at the population structure of wild barley and the cultivated landraces with a view to answering this question.

Barley geneticists have tried to standardize their efforts by setting up what is called the Wild Barley Diversity Collection (WBDC). This collection is made up of 318 wild barley strains (accessions), selected both to represent the broadest possible array of non-landrace strains, and to represent as much as possible of the ecological diversity within which barley flourishes. Most accessions are from the Fertile Crescent, the area of the Near East where most scientists think barley was first domesticated, but some are from Central Asia, North Africa, and the Caucasus region between the Black and Caspian seas. The landrace counterpart collection of barley used for comparison is from a center called ICARDA (International Center for Agricultural Research in the Dry Areas), which contains 304 worldwide accessions. Some studies use this collection exclusively, but others also include a broader sampling of cultivated strains in order to cover as much geographic and genetic diversity as possible.

To make analysis of the genomes of these many strains easier, researchers exploited certain reproductive characteristics of the barley plant. Individuals of barley and other grains can mate with themselves, and indeed have found that this is the best way to reproduce. They breed with other individuals occasionally, but their preferred mode of reproduction is with themselves. This selfing mode of reproduction means that they behave a little—but not exactly—like clones of themselves. It also makes it easier to trace their genetics and to reconstruct their origins than it would be with a sexually reproducing species like ours—for as we all know, sex complicates everything. To make the barley study as easy as possible, the accessions used were forced to reproduce with themselves for three generations before being harvested and processed.

Several groups of researchers examined the genetic dispositions of varieties within the species *Hordeum vulgare*. Joanne Russell, Martin Mascher, and colleagues looked at the barley landraces using a technique called whole exome sequencing. This technique obtains genome sequences from regions of the genome that code for proteins. There are several million data points in each of these genome surveys, and making sense of all that data is a big informatics problem, to which we explored available solutions in Chapter 5.

Figure 7.3 shows a principal components analysis of over 250 individuals of *H. v. vulgare* and the wild *H. v. spontaneum*. It emerges from this analysis that all the landraces are more similar to each other than they are to the wild strains (*H. v. spontaneum*). Although there are many issues with this kind of approach, the diagram still offers us a picture of how the individual landraces and wild strains might be related to each other—or at the very least, new pathways to thinking about how these strains are related.

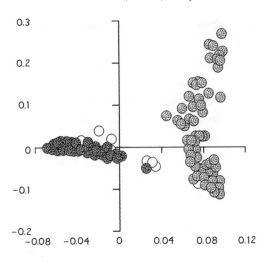

Figure 7.3. Principal component analysis of landraces of *H. vulgare* (dark circles) and of wild strains (*H. v. spontaneum;* gray circles). Each dot represents one of the more than 250 individuals in the study. Open circles are strains that were originally classified as wild *H. v. spontaneum,* but appear to be closer to cultivated landraces. The X axis represents the sequences that explain the highest proportion of the data, and the Y axis represents the next highest amount of variation explained. The values along the axes are arbitrary. Adapted from Russell et al. (2016).

Ana Poets, Zhou Fang, Michael Clegg, and Peter Morrell examined a larger collection of barley landraces (803 of them) to see if there is any clustering within the landraces. And they found a lot of it, with six major clusters (Figure 7.4). More surprisingly, in two-dimensional space those clusters can be overlain on a map of where the landraces are found. For instance, the dark gray landraces that cluster together in the diagram below are from the Fertile Crescent, and the light gray strains are found in Central Asia.

These studies are interesting because they indicate that landraces tend to stick to certain geographic regions. As Poets and her colleagues observe, "Despite extensive human movement and admixture of barley landraces since domestication, individual landrace genomes indicate a pattern of shared ancestry with geographically proximate wild barley populations."

Such research can also help us estimate the number of clusters,

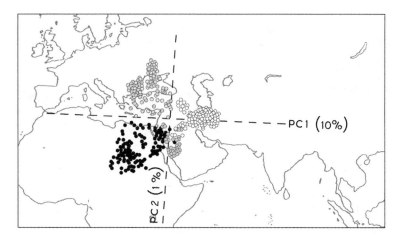

Figure 7.4. Principal component analyses of barley landraces (*left*), and wild barley (*right*). This PCA was generated using 803 landraces, overlain on a map of the region where the landraces are found and assuming four clusters (adapted from Poets et al. [2015]). The different shades in the circles represent the four different assumed clusters: one from central Europe, one from the coastal Mediterranean region, one from east Africa, and one from Asia.

or populations, of barley landraces and wild strains. The estimate for wild barley and its landraces ranges from four to ten. The number of clusters is a little fuzzy because trying to discern the number of clusters from PCA analyses is very subjective. Try it yourself, with the mapped data in Figure 7.4. Ignore the different shades, and try to draw circles around what you think are clusters. Some readers might end up having more than ten circles in their map, or as few as two.

As we saw in Chapter 5, STRUCTURE plots can give a detailed picture of the clusters or populations in a data set. We will discuss two of them here. The first is from the Poets and colleagues study (Figure 7.5). They assumed a K of 4 (that is, that there were four ancestral populations—central European, coastal Mediterranean, east African, and Asian). The approach visually shows four populations, but you will notice that there is a lot of uncertainty here, as shown by the leaking of shades among some individuals. The implication is that, while there do appear to be four structured populations, there also has been significant admixture among the landraces.

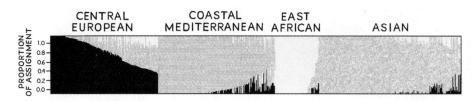

Figure 7.5. STRUCTURE analysis of 803 barley landraces, with K=4. These are the same four clusters in the PCA analysis of Figure 7.4: one from central Europe, one from the coastal Mediterranean region, one from east Africa, and one from Asia.

The second study, by Joanne Russell, Martin Mascher, and colleagues, included 91 wild and 176 landraces from the Fertile Crescent of the Near East. The scientists narrowed the geographic range of their analysis because they were primarily interested in the genetics of five special accessions. They separated the landrace individuals from the wild accessions, and set the number of ancestral populations at five (K=5). Their wild accessions were then assigned to these five ancestral populations. The wild strains fall into two recognizable clusters, suggesting that they come from two well-defined ancestral populations (Figure 7.6). The geographic break between the two clusters appears to be between a group of accessions mostly from Israel, Cyprus, Lebanon, and Syria, and those from Turkey and Iran.

Once they had obtained the detailed picture of wild strains, the researchers analyzed the landraces as shown in Figure 7.6. This diagram gives a good picture of how different the wild accessions are from the domestic landraces. Russell, Mascher, and colleagues suggest that there are three clusters. These may or may not visually pop out at you, and they have very little of the wild population variation in them. The analysis tells us, though, that there are at least three ancestral patterns for landrace barley from this region. The five special accessions mentioned earlier are included in this analysis; and they are as special as accessions get, since they consist of six-thousand-year-old barley kernels, found in Israel, that are believed to represent cultivars that humans used all that time ago. And they appear to be very similar to modern landraces. More specifically, these cultivars show close affinity to current landraces from Israel and Egypt. This result is spot on with the

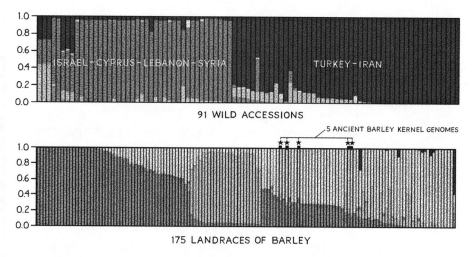

Figure 7.6. *Bottom:* STRUCTURE-like analysis of 91 wild accessions used in the Russell et al. study. The number of ancestral populations was set at K=5, and the different shades reflect the assignment of the individuals in the analysis to these five populations. There are two very visually evident clusters. *Top:* STRUCTURE-like analysis of 176 landraces of barley using K=5 for the number of ancestral populations. The shades of the bars represent the same ancestral populations as the bottom part of the figure. Stars indicate the five ancient barley kernel genomes. Adapted from Russell et al. (2016) and Mascher et al. (2016).

idea that the domestication of barley was initiated in the Upper Jordan Valley. Close examination of the ancestral components of these five samples (marked in black in Figure 7.6) suggests that the Israeli landraces grown today have not changed much in six thousand years, despite some occasional mating with wild strains.

Genome-level information is instructive not only about the ancestry of barley, but also about the genes that might have been involved in its domestication. We have already discussed the major outward difference that distinguishes wild accessions from landraces—the brittle spike. But other traits were certainly also selected for by barley breeders over the past ten thousand years. Indeed, Russell, Mascher, and colleagues used their data set to identify the kinds of genes that have been, and continue to be, under selection in landraces. Among the traits they showed to be under breeding selection over the past several millennia are days to flowering, and height as response to temperature and dryness. Both traits are important in the adaptation of cultivated barleys to their domestic circumstances. But as the scientists point out, there are doubtless many factors still to be uncovered. More genomics work will help us work out what they are.

What about the brittle spike trait that we have seen was perhaps the most important genetic change during domestication? It turns out that the brittle rachis trait is under quite simple genetic control. Two genes are involved, Btr1 and Btr2, whose protein products interact with each other. When these two gene products interact properly, the rachis is brittle; but if there is an abnormal interaction as the result of gene mutation, the rachis stays strong, and no shattering occurs. Other domestic grains, such as rice and wheat, also have strong rachises, raising the question of whether breeders of rice, wheat, and barley selected for this trait in these grains via the same genetic pathways. Mohammad Pourkheirandish and Takao Komatsuda settled this question by showing that the brittle rachis trait in barley is in fact unique: the rice and wheat systems do not involve the Btr1/Btr2 interaction. Clearly, there is more than one way to achieve the same rachis qualities. This is a common theme in evolutionary biology, so it is hardly surprising that plant breeders have also stumbled onto the same principle using artificial selection.

In the first sentence of his 2015 review of barley biology, Robin G. Allaby summed up the understanding of the domestication history of barley in eight words: "Barley did not come from any one place." This shrewd observation is important, because most researchers have long assumed that domestication is necessarily a singular event. Allaby clarifies our interpretation of the genomic data by pointing out that every single landrace of barley so far examined has genomic remnants of the four or five ancestral wild accessions, and he raises a key question—is barley the exception among domesticated forms, or is it the rule? The answer is that barley might well illustrate the rule. Domestication—

which in the case of barley seems to have taken place over the general region of the Fertile Crescent—was evidently not a simple process.

In the past, the breeding of landraces of barley possessing the most desirable traits for agriculture was a trial-and-error affair. Six thousand years ago, barley farmers knew nothing of formal genetics, but they were smart and clearly knew enough about their plants to achieve the results they wanted. Breeders continue to grapple with the same two major kinds of traits: yield and quality. Yield traits include features like numbers of seeds set, capacity to breed multiple times a year, or the brittle seed character that, if mutated, allows for more efficient harvesting. Quality traits are those that impact the protein content, oil content, or any other phenotype concerned with the nutritive content of the plant. During the twentieth century, barley breeders were still using their knowledge of classical genetics to facilitate breeding in a tedious and labor-intensive process. With the rise of genomic technology, and the ease with which it can be applied to large numbers of lines and landraces, a very different approach to barley and other grain breeding has now become possible, using cheaper and faster techniques.

Genome-based plant breeding uses a concept called genomic prediction that relies on the predictive abilities of traits. It requires genome-level sequencing of large numbers of landraces, as well as abundant data on the traits that might be targeted (such as seed size, protein content, and protein yield). Prior to the use of this approach, barley breeding experiments were massive and costly. Now, using genomic prediction, barley breeders can get a more precise, quicker, and cheaper idea of how easy it will be to breed for certain traits. Several such studies have already been directed at the assessment of quality traits that are important in brewing.

Malthe Schmidt and colleagues analyzed the predictive abilities of twelve malting characteristics of spring and winter barleys. By ranking those twelve desirable malting traits, they showed that winter barley would be easier to work with. Another study demonstrated the feasibility of the genomic remnants in improving seed quality traits. Nanna Nielsen and colleagues examined features such as seed weight, protein content, protein yield, and ergosterol levels (generally thought to be an indicator of resistance to fungi and bacteria), showing how genomics could predict the efficacy of breeding programs for these traits too. So, while it is still early days, genomic approaches have already demonstrated their ability to facilitate improvement in the efficiency, yield, and quality of barley cultivation. Still, it is quite likely that the future of barley will lie in an even more cutting-edge technique: direct "gene editing" using the brand-new CRISPR technology that has received so much recent publicity. However the story plays out, one thing is certain: molecular biology holds huge promise for improving the raw materials of maltsters and brewers.

8

Yeast

The slender, shiny brown bottle bore no label, but close scrutiny of the almost illegible letters on a raised glass ring around its neck revealed the words "Trappisten Bier." The crown cap was more specific: "Trappist Westvleteren 12, 10.2%." Our initial shock over actually holding a bottle of the world's most legendary beer was rapidly replaced by feelings of intimidation. Brewed in tiny quantities by the monks of the Saint-Sixtus Abbey in Flanders, and normally available only at the abbey itself under cloak-and-dagger circumstances, the fluid inside this bottle had regularly been voted "world's best beer," a deep, nutty brew with vivid yeast flavors often ascribed to an unusually high live yeast content. Eventually, we screwed up enough courage to remove the cap. The world's best beer? Well, on a planet in which the most delightful thing about beer is its sheer variety, that's an extremely tough call. Let's just say that the beautifully harmonious contents of the bottle did not disappoint.

We are literally swimming in a sea of microbes, every hour of every day. The number of different microbe species that live within and upon us is estimated at about ten thousand—two to three times the number of plant species thought to exist in a typical rainforest, and about the same number as all the bird species on our planet. And this is just the microbial life that "sticks" to us. No wonder our late colleague Stephen J. Gould declared that there was never an age of dinosaurs or an age of man. Rather, we have all always lived in the age of microbes.

Not every human has the same microbes residing in and on him or her. Every area of your body hosts a different community of microbes. These tiny unicellular creatures belong to one of three great groups or domains: Bacteria, Archaea, and Eukaryota. All three are descended from the single common ancestor that links all life on Earth together, as we can determine from comparing the genomes that carry their reproductive blueprints (Chapter 5). Members of Bacteria and Archaea are strictly single-celled organisms that do not have a nuclear membrane around their genomes, whereas eukaryotes can be single celled or multi-celled, and have a walled-off nucleus. Like human beings, the barley and hops that go into beer are multi-celled eukaryotes. But the third great ingredient of beer, the yeasts, are single-celled eukaryotes.

The yeasts are part of the major Eukaryote group called fungi, which also includes the mushrooms. Believe it or not, mushrooms are not single organisms but rather organized colonies of single-celled organisms all belonging to the same species. Mushrooms have very familiar forms and morphologies that make them relatively easy to classify. Largely because of their nondescript anatomy, the tiny yeasts are more difficult to classify by eye, even using powerful microscopes. Still, despite their simplicity of form, the variety of lifestyles these simple creatures can adopt is stunning. And it has led to a vast array of species and evolutionary patterns. We need look no farther than our own everyday lives to verify this observation. Hardly a day passes in which we don't eat a dish produced using a fungal species such as one of the many mushrooms. Fungi can also be the source of some of our most stubborn and uncomfortable illnesses, as well as of many minor ailments such as athlete's foot. And for some of us fungi might even have been the source of mind-expanding experiences: the psilocybin compounds found in over 150 species of fungi are famous for their psychedelic effects. Oddly enough, all fungi are more closely related to animals than to plants. When a vegan eats a salad with mushrooms, he or she is arguably cheating.

There are two major kinds of fungi, plus a few stragglers that are so different that they are placed in their own major groups. One of the major groups, Basidiomycota (which encompasses puffballs, mushrooms, and stinkhorns, for example) is probably more familiar to most people, but the second major group, Ascomycota, contains the species important to beer, bread, and wine. A large collaborative group of researchers led by Rytas Vilgalys at Duke University studied two hundred of the better-known species of fungi to determine how they are related to each other, using DNA sequence information to construct a genealogical tree in the manner we detail in Chapter 14. This tree fortunately confirmed much of what was already believed about fungal relationships, but it also indicated the position of several new kinds of fungi for the first time. It highlighted how little we know: although there are by now around one hundred thousand formally described species of fungi, some researchers suggest that there may be between 1.5 million and 5 million fungal species on our planet.

Although the major player in the making of beer, bread, and wine is the ascomycete *Saccharomyces cerevisiae*, also known as brewers' yeast, several other fungi also affect brewing, in both desirable and undesirable ways. As with the other organismal components of beer brewing (barley and hops), an understanding of the genetic or genomic makeup of yeast is becoming steadily more important as the science of brewing advances. There is still a tug of war between traditional approaches and the more recent genomic technologies, but for the most part brewers are pretty open to using the information that genomics affords their craft.

Saccharomyces cerevisiae was one of the first eukaryotic organisms to have its genome sequenced, in 1996. When whole-genome sequencing emerged in the 1990s, this yeast species was an obvious candidate for sequencing, both because it is economically important and because its genome is small (twelve million bases, as opposed to the three billion bases in our human genome). We estimate that its initial sequencing probably cost the consortium that did it on the order of ten to twenty-five million dollars: a huge figure due not only to the many unknowns involved, but also to the clunky and costly nature of the first-generation sequencing technologies then available.

As a result, by 2005 only a handful of yeast species had been examined for whole-genome variation. But now a hundred yeast genomes can be sequenced in less than a day, and for a fraction of what the first yeast genome sequence cost (probably no more than a hundred dollars per genome). This dramatic change has occurred for two reasons. First, once a genome in a major group has been generated it can serve as a scaffold, or reference, for other genomes of related species. Second, sequencing has morphed into what is called next-generation sequencing, and even into next-next-generation sequencing. To illustrate how much acceleration has occurred, a postgraduate student of the 1980s might have devoted an entire thesis to sequencing a single gene from a single species. In the 1990s a comparable research project might have expanded to hundreds of thousands of bases, and to several species. But students in the 2000s could easily do tens of millions of bases for a hundred or so species; and, by about halfway through the current decade, technological advances had allowed this number to jump to hundreds of millions—if not billions—of bases. Today a student can routinely generate up to thirty billion bases of sequence, and just one student can do all the work that was accomplished in genomics theses during the 1980s and 1990s in less than a second, at a tiny fraction of the cost.

Given these advances, it is hardly surprising that researchers have now analyzed several thousand yeast strains and species in their quest to discover which wild yeast species is the closest relative to those essential for making beer, bread, and wine. The scientists who work on this problem call those domestic strains captive yeasts, and an aid in their search has been the existence of centralized repositories for *Saccharomyces cerevisiae* strains and their close relatives. One of the largest of these is at the Insti-

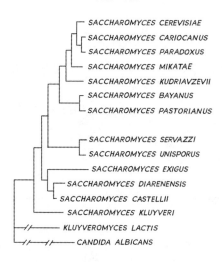

Figure 8.1. Phylogeny of *Saccharomyces cerevisiae* and closely related species. The branch lengths are proportional to the amounts of change accumulated by the species. Adapted from Cliften et al. (2003).

tute of Food Resources in Norwich, Great Britain, which contains more than four thousand strains.

The yeasts involved in beer-, bread-, and wine-making come primarily from the single family Saccharomycetaceae. This family contains thousands of species; but, as noted, *Saccharomyces cerevisiae* is the one that is essential to the production of these commodities. The history of *S. cerevisiae* and its close relatives is both interesting and convoluted. Figure 8.1 shows the relationships among these species, although it appears we are still looking at a moving target, made no easier by hybridization among species. One odd species not in the figure is *S. eubayanus*, which grows at low temperatures and, along with *S. cerevisiae*, is one of the parents of the lager yeast *S. pastorianus*. Note also that there is also a species in the group called *bayanus*. Whenever taxonomists add a prefix to an already known species it means something specific, and "eu" in this taxonomic context means "true." But *S. bayanus*, also used in winemaking, has actually turned out to be a three-way hybrid of *S. cerevisiae*, *S. uvarum*, and *S. eubayanus*. The last of these was actually discovered after *S. bayanus* had been brought to light, making it a good example

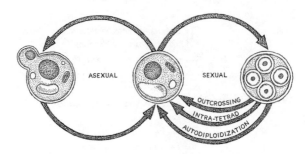

Figure 8.2. The life cycle of yeast. When nutrient content is high, the yeast cell in the middle is happy to reproduce asexually (cycle on the left). It produces a "schmoo" (*far left*), and buds off a sister yeast that then enters its own cycle. If, on the other hand, nutrients are scarce, the yeast cell "decides" to reproduce sexually. Its genome will produce gametes or tetrads (*far right*). It then has several options for sexual reproduction, in one of which one cell of its tetrad meets a single cell from the tetrad of another individual. There is a complex mating structure for the sexual reproduction cycle.

of how confusing yeast systematics can be. It took modern genomics to tease apart the complexities of the story.

For most of its life *Saccharomyces cerevisiae* lives like a devout monk, refraining from sex. But there are times when it can become amazingly promiscuous. The diverse lifestyle of this species of yeast varies according to how happy the yeast population is when the time comes to reproduce. By happy, we mean how many nutrients are available to it. Figure 8.2 shows the life cycle of this budding yeast. When times are good, yeasts will reproduce asexually; but when times are tough, and there are few nutrients around, beer yeasts will go sexual and make spores. Here we see an echo of the closer relationship of fungi to us than to plants. Plants can use both sunlight and nutrients from the soil to produce the energy needed for their daily lives. But, like us, fungi need nutrients such as carbohydrates, and a shortage of these forces a change in the sexual strategy of *S. cerevisiae*. Instead of budding off the genetically identical daughter cells that yeast biologists affectionately call "schmoos," the yeast produce haploid spores, equivalent to our sex cells, which provide them with a mechanism for exchanging genetic material with other yeasts (Figure 8.2)—and for the occasional creation of a hybrid yeast. Usually, though, there are plenty of nutrients around and *S. cerevisiae* are abundant in the environment—a quality that has made them a favorite study subject for scientists, who find them easy to grow in the lab and a very useful model for understanding the interactions of proteins and how they are controlled by genes.

The laboratory approach to determining the progenitor of beer yeasts resembled that used in the case of barley (Chapter 7), with the closest wild species and subspecies used to anchor the search. The species *Saccharomyces paradoxus* was chosen for this role because it has apparently escaped captivity and is not used as a domestic yeast. It can thus serve as a model of what *S. cerevisiae* might have been like if it had not been "captured." Having made this choice, the researchers examined the geographic population structures of beer yeast strains against a background of nonbeer yeasts, including wine and sake yeasts, medical samples, and yeasts taken from natural sources such as fruit or tree exudates. The population structure of *S. paradoxus* emerged as quite clear, with distinct genomic boundaries between the various geographic regions where the species is found. The STRUCTURE analysis revealed four very distinct populations: one from Europe, one from East Asia, one from North America, and one from Hawaii. Specifically, European, Far Eastern, and American strains can be diagnosed as such with 100 percent certainty, and the strain from Hawaii appears to be about 80 percent Hawaiian and 20 percent North American. The distinctness of the non-captive yeast populations is presumably due to a lack of manipulation by yeast biologists and brewers.

Gianni Liti and colleagues arrived at very different results when they examined the genomics of thirty-six strains of *Saccharomyces cerevisiae* strains, including wine-making, clinical, and baking yeasts. Here they found the assignment of individuals to ancestral strains very difficult, and although most of the strains they used were wine yeasts, not terribly relevant to brewing, they were able to document that sake and wine and beer yeasts show a well-defined separation, indicating that they might have been kept separate since each was first used to ferment a beverage (though remember Ninkasi and the bread that probably went into her beer). This also suggests that separate instances of human ingenuity (or luck) had resulted in the capturing of the yeast strain involved. In a much larger study, conducted in 2016 and includ-

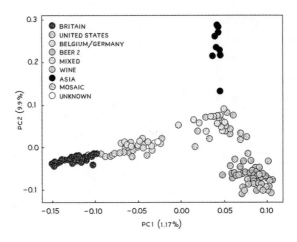

Figure 8.3. Principal component analysis of the *Saccharomyces cerevisiae* genomic data from the Maere/Verstrepen group. Adapted from Gallone et al. (2016).

ing 157 fermentation strains of *S. cerevisiae,* Kevin J. Verstrepen and colleagues more precisely pinned down the genomics of brewing yeasts. Let's examine the structure of their captive yeast strains in detail by taking a step-by-step look at what the data tell us, using the genomics tools we discussed in Chapter 5.

First, the Verstrepen group sequenced its 157 strains *de novo,* meaning they used classical genome sequencing methods, not targeted sequencing. This approach was possible because of the small size of the yeast genome, which also allowed the genome sequences of all the strains to be of extremely high quality because a median of 675 million base pairs could be sequenced for each strain. Recall the importance of coverage, which simply means the amount of DNA sequenced divided by the size of a single genome of the organism. In this case we have an average of 675 million divided by 5 million, giving us about sixty-eight-fold coverage of each strain. This degree of coverage is impressive, and it ensures that few if any sequencing errors exist in this data set.

Our first approach to understanding this massive amount of data is to look at the information using a principal components analysis (Figure 8.3). With our untrained eyes we see four clusters—maybe five, or maybe three, but four seems a good number on which to settle. Note

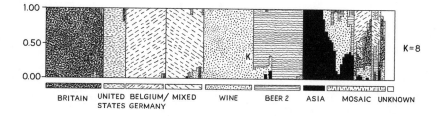

Figure 8.4. STRUCTURE analysis of 157 yeast strains. The key for the populations is given at the bottom of the graph, and indicates populations of origin as in Figure 8.3. Adapted from Gallone et al. (2016).

that the plot accounts for about 20 percent of all the variability among these strains, meaning that there is still a lot of information being left out of the analysis. What that analysis has achieved is to reduce hundreds of dimensions to two, to make visualization easier. The analysis is coarse, but it does appear that beer yeasts pop up in two areas of the space. One is the horizontal "streak" of strains, and the other is the cluster with the wine yeast strains (lower right hand of the graph). The vertical streak represents Asian sake yeasts that were already known to be quite different from other *S. cerevisiae* strains.

After setting the structure at eight populations (K=8), the number of populations deemed the most likely using statistical tests, a more rigorous STRUCTURE analysis of the yeast strains gives us Figure 8.4. Several geographic regions can be identified with very specific populations, as indicated by the solid blocks in the diagram. Oddly enough, it appears that the beer yeasts labeled "Beer 2" have some affinity to wine yeasts. The Beer 2 group is an odd mixture of yeast strains from Belgium, the United Kingdom, the United States, Germany, and eastern Europe. Mixed yeast strains appear distinct, but also have elements of several yeast populations. Mosaic strains are aptly named, as they appear to be mishmashes of all the various populations in the analysis.

The potential hierarchical relationships of these yeasts with K=8 is not clearly evident. Lower numbers of populations tend to lump some of the geographic designations together, but the best way to examine the potential hierarchical relationships is to perform a phylogenetic analysis (Figure 8.5). Note that the several wild strains of yeast are at

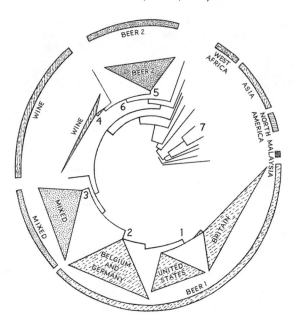

Figure 8.5. Phylogenetic analysis of 157 yeast strains. Node numbers are explained in the text. The lengths of the branches, and the depths of the group triangles, represent the amount of change in the strains concerned. The categories assigned by Gallone et al. (2016) are in the outer circle of the diagram. Adapted from Gallone et al. (2016).

the base of the tree—by definition, because the captive yeasts all had to derive from the wild yeasts with which the tree was rooted. The topology of the tree and the position of the many industrial strains within it also suggest, as Verstrepen and his colleagues observe, that "the thousands of industrial yeasts that are available today seem to stem from only a few ancestral strains that made their way into food fermentations and subsequently evolved into separate lineages, each used for specific industrial applications."

The tree itself demonstrates several important aspects of beer yeast history. First, it indicates that the British and Belgian/German beer yeast groups share a single ancestor (see Node 2 in Figure 8.5). This means that the yeasts used in these two European areas have been kept reasonably separate from other yeasts. If this clean separation had not been part of their ancestry, we would see strains from other geographic origins invading the British and/or the Belgian/German parts of the tree. It also appears that United States yeast strains are more closely related to the British yeast strains, as shown by the connection of yeasts from these together to the exclusion of Belgian/German yeasts (Node 1). Mixed yeast strains are indeed aptly named, because the ancestor from which all the mixed strains arise (Node 3) gives rise to a plethora of strains from different geographic areas that also include the bread yeasts. The wine yeast group isn't entirely pure with respect to its position in the phylogenetic tree, as several beer yeasts and others crop up in the same group as the wine yeasts (Node 4). The yeasts of the Beer 2 group, while composed of a mixture from Belgium, the United Kingdom, the United States, Germany, and eastern Europe, do all come from a single common ancestor (Node 5). As we suggested earlier, the Beer 2 group of yeasts has some affinity to wine yeasts, and the phylogenetic tree strongly confirms this (Node 6). Finally, sake yeasts are also derived from a single ancestor, and their position in the tree (Node 7) suggests that they are ancestral to all other industrial yeast strains.

At this point, those interested in brewing should be wondering about lager yeasts versus ale yeasts. They should be quite different—and hence located in different parts of the tree—because it is well known that lager yeasts tend to do most of their work on the bottom of the fermentation vat, while ale yeasts ferment above, leaving a thick residue at the top of the vat. In addition, and probably more importantly, ale yeasts work best at close to room temperature, while lager yeasts work better at significantly colder temperatures. What's more, Joanna Berlowska, Dorota Kregiel, and Katarzyna Rajkowska clearly showed in 2015 that the genomic and physiological properties of lager yeasts are very different from those of ale yeasts, so we should expect them to be quite separate from each other.

This is where it gets complicated. Until recently, all lager yeasts were considered to belong to the species *Saccharomyces carlsbergensis*, which, as we've seen, has now been identified as an interspecific cross between the common beer yeast *S. cerevisiae* and the closely related *S. eubayanus*. Before this hybridization event, though, one of the ancestors had evidently duplicated its genome. When these events happened is unknown (though it was probably more than five hundred years ago—see Chapter 2), but genome duplication and hybridization are shocking events in any lineage. To confuse matters further, the hybrid lager yeast has been given the name *S. pastorianus* and, as noted previously, there is even some controversy over the involvement in the beer yeast business of yet another species, *S. uvarum*. But wouldn't we expect all these complicating factors to shore up the idea that lager and ale yeasts should be in different parts of the yeast tree? Not necessarily. Several different lager yeasts, from the United States, Germany/Belgium, and Beer 2 groups, were included in Verstrepen and colleagues' study, and they turned out to be distributed throughout both the Beer 2 group and the Germany/Belgium group. This may appear bizarre, because it flies in the face of the big differences between the two kinds of beer yeast. But then again, beer yeasts are found in the wine yeast group, and there are two distinct beer groups. Apparently, in the world of captive yeasts anything goes.

Although we have focused on yeasts within the genus *Saccharomyces*, other genera of yeast are used—with various degrees of trepidation and adventurousness—in brewing beers. For the most part, non-*Saccharomyces* yeast species have historically been looked upon as nuisances, and as potential skunkers of brews. Forty years ago, an early batch of home brew that one of us created was not carefully enough cooked and fermented, so that while it came out with a relatively good alcohol content, it was hazy and odd-tasting. An unknown yeast species, presumably wild, had infected the wort and taken over from the ale yeast that was supposed to do the fermenting. This was an accident, but with the recent rise of sour beers, saisons, and farmhouse ales, other species of yeast not very closely related to *Saccharomyces cerevisiae* have come into the mainstream brewing picture. Two genera of yeast are incredibly important in brewing these styles of beer—*Dekkera* and *Brettanomyces*. Although most of the non-*Saccharomyces* species used in brewing are relatively closely related to *Saccharomyces*, these two genera, along with another genus called *Pichia*, are quite far away on the tree. Evidently, because of the extreme diversity of yeast genera, there are many potential kinds of yeast available to experiment with. Better to experiment than to contaminate, we would guess; but then again, a lot of the history of beer brewing has involved serendipity.

As if all these discoveries about beer yeasts weren't enough, there are still several very important inferences to be made from the Verstrepen group's impressive study. Some of them involve using genome biology to manipulate yeasts genetically (see Chapter 16). But others are important for understanding the biology of beer yeasts. First, look at the tree closely. The branches of the tree are not of uniform length, because by design the lengths represent how much change has occurred along a lineage. Now look at the wine strains, and at any of the beer strains, and notice that the wine yeast branches are stubbier than the beer yeast branches. This means that, over a similar period, the genomes of the beer yeasts have changed more than those of the wine yeasts.

Anthony R. Borneman and colleagues looked closely at the genomic diversity of 119 wine strains in 2016, and noticed that wine yeasts were very homogeneous: their genetic variability is far less than they had expected. In fact, their analysis indicated a bottleneck in genetic diversity. Think of it this way. You have a bag of white, red, and black marbles, in equal numbers, and place them in an old-fashioned milk bottle with a tapered neck. You shake them up and try to pour them out. Inevitably only a few will emerge, while the rest will lodge in the neck of the bottle. If only a few pour out, the 1:1:1 ratio of colors will be eliminated, and the population will look very different after the pouring. Indeed, it might very well happen that only one color comes out.

If we think of the marbles as representing genes, and the colors as alleles, we have here a good metaphor for genetic bottlenecking. Once a bottleneck has emerged, inbreeding will inevitably follow, reinforcing the new and probably impoverished pattern of variation. It is this phenomenon that evidently established the very short tree branches for wine yeasts. Domesticated beer yeasts, in contrast, show a lot more genomic variation. When beers are brewed, the yeasts in the brew are not exposed to the long periods of nutrient starvation that afflict wine yeasts. As we saw, yeasts will reproduce quite happily without sex (Figure 8.2), and starvation is necessary to induce sporulation and to push them into reproducing sexually. Many beer yeast strains consequently have a diminished capacity to sporulate, and some have even completely lost the capacity altogether. Indeed, most of the yeast strains in the Beer 1 group cannot sporulate. This capacity to avoid sex, or even to lose the capacity for it, is characteristic of domesticated strains. For a wild yeast in an unpredictable environment this would be a risky strategy, but it was apparently one that brewers forced on many beer yeast strains.

And in fact, it turns out that avoiding sex makes great sense for beer yeasts. Brewers' yeasts are typically used in one cycle of brewing, and are then transferred or recycled for the next, and so on. Since brewers usually go from one batch to the next quite rapidly, there is usually no prolonged storage period, and the yeast are usually happy and well fed. This contrasts with the seasonality of wine-making. Wine yeasts are happy for only a short period each year, when they are carousing in the bubbling must. The rest of the year they cling on in dried-out barrels, in the vineyards, and even in the guts of insects. During these tough times, with a reduced probability of making it to the next fermentation cycle, the wine yeasts will routinely go sexual, and live outside their fermentation lifestyle. Most of the time, then, the wine yeasts will have a very small population size compared to those beer yeasts, which leads to three very interesting population effects.

First, because of the population size differences, beer yeasts will evolve more rapidly and change more than wine yeasts, a phenomenon that is borne out by the results in Figure 8.5, and can also be seen in their much greater genetic variability. Second, because brewers are somewhat proprietary, when they find a good combination of ingredients they tend to keep it to themselves. This results in the isolation of beer yeast strains from one another, increasing their differentiation. Finally, since beer yeast strains show a general loss of capacity for sexual reproduction, and because they are relatively happy (with no extremes of natural selection forced on them), they can tolerate more mutations (hence variation) in their genomes.

Yet if a brewer does not provide a good environment for it, the captive beer yeast will do poorly, so that the last several millennia of beer brewing have been a big forced evolutionary experiment for them.

Some yeasts have remained wild and have maintained a lot of genetic variability. Others have been captured—domesticated—and now behave very differently from their wild progenitors. And because the captive niches they live in are highly constrained, still others have evolved very specific responses to their circumstances. Along the way, beer yeasts seem to have acquired two characteristics of domesticated organisms: extreme genome specialization, and extreme niche specialization. Fortunately enough variation remains, both inside the genus *Saccharomyces* and outside it, to ensure that beer yeast biology will remain interesting for as long as we brew beer.

Finally, the latest wrinkle in the saga of yeast and beer involves a radical departure from tradition. Since the beginning, brewers have been restricted to making beer in batches. After fermentation is complete and the beer has been drawn off and bottled, the brewing equipment must be cleaned of both dead and living yeasts so that the process can start all over again from scratch. But what if beer could be produced continuously, much as many spirits are nowadays? The University of Washington chemist Alshakim Nelson has proposed a way of doing this. Using three-dimensional printing techniques, his team has produced minute hydrogel bioreactors in which a population of yeast can flourish and be active for months at a time. When these tiny yeast-infused cubes are dropped into a glucose solution they set to work doing what yeast do so well—fermenting it, in a process that continues for as long as the solution is replenished. Why the yeast abandons its life-and-death cycle under these conditions is not yet known, but the possible implications of this new approach for the future of brewing are intriguing, to say the least.

9

Hops

Those 2,600 international bitterness unit (IBU) hop bombs may be out there, but they're not easy to find and probably for good reason. Our own ransacking of Manhattan Island beer shops didn't yield anything more extreme than a Triple IPA with 131 IBU. Three bold hop cones dominated its label, making us wonder about the "Not Meant for Aging" injunction printed just above them. After the cap came off, we were hit with a nose redolent of bitter Simcoe hops; but on the palate the aromatics and the 11.25 percent alcohol took over, resulting in a sweet, fruity, and almost mellow ale, backed up—but not overwhelmed—by the abundant hops. The balance was beautiful; and to be honest, we were quite content not to have found anything with a higher IBU.

Two plants provide the core ingredients for modern beer: barley grains, and the dried seed cones of the hops vine, *Humulus lupulus*. Although barley was in there from the beginning, the addition of hops was something of an educated afterthought: while the history of barley and beer go hand in hand back at least to the beginnings of human settled life, the routine inclusion of hops in beer goes back barely a millennium (Chapters 2 and 3). Traditionally, European beer makers had flavored their product with a gruit of wild herbs, but in the ninth century these herbal concoctions began to be replaced by the addition of hops. With this change came multiple advantages, because hops not only added a refreshing bitterness, but also acted as a preservative. Not everyone adopted hops right away, though: the British converted only slowly, in a process not completed until the sixteenth century.

One reason for this late adoption may have been the rather dubious reputation that came along with hops. In the twelfth century the abbess Hildegard of Bingen lamented that hops "make melancholy grow in man and makes the soul of man sad, and weighs down his inner organs." On a more profane level, far back into the mists of time hops have been blamed for the two unfortunate male conditions now inelegantly known as brewer's droop and man boobs, both conceivably—but not yet provably—related to the plant estrogens that hops are now known to contain. Yet on their own, the dried seed cones of the hops plant were also widely used in medieval medicine, among other things for the treatment of toothaches and kidney stones. They were also prized for their calming effects, and until not too long ago they were routinely used to stuff pillows for a better night's sleep.

Politics may have played a role, too. According to one sixteenth-century ditty,

> Hops, reformation, bays and beer
> Came to England all in one year.

The reference here is to the coincidence between the introduction of Protestantism into England and the rise of hopped beer during the reign of Henry VIII. With the political wind behind it, the hops plant made it big in England from the mid-sixteenth century on, and few have regretted the change. Also notable is the cessation of monastic production of gruit beers that began in Germany just before the turn of the sixteenth century, due to the regional adoption of the *Reinheitsgebot* purity laws. Although the ostensible intent of these laws was to control the use of different kinds of grains to keep bread affordable, they also had the important political effect of weakening the power of the Roman Catholic church. Gruit beers, it seems, were a drink that the religious reformers wanted to eliminate.

The purity laws also affected the development of beer-making techniques. For half a millennium, the push to standardize beer to its three then-known ingredients—water, barley, and hops—basically limited experimentation to barley and hops alone. Fortunately, all that has changed in the past few decades, which have in many ways been a return to the beginning. But although many brewers nowadays are wildly experimenting with all kinds of parameters in beer-making, hops remain a crucially important ingredient in beer. So let's take a closer look at this remarkable vine.

Like wheat and barley, hops are flowering plants. But while barley is a monocot, *Humulus lupulus* is a member of the other grand lineage of flowering plants, the dicotyledons (dicots). The big distinguishing characteristic between the monocots and the dicots concerns the cotyledon, a simple sheath of tissue that generally develops into a plant's first leaves (Chapter 7). Basically, the difference is that in monocots there is one of these sheaths, and in dicots there are two. Figure 9.1 summarizes the classification of hops among the plants.

The more than two hundred thousand species of dicotyledons are grouped by botanists into smaller taxonomic units, based both on their anatomies and on molecular information. The first split in the dicots is between the eudicots and a tiny, strange group of plants commonly known as hornworts. The next split is between several other odd lineages and the "core eudicots" to which hops belong. The core is divided into the rosids (including hops) and the asterids (which include lots of food plants such as eggplants, potatoes, peppers, common sunflower, tomatoes, coffee, and many common table herbs). The rosids are di-

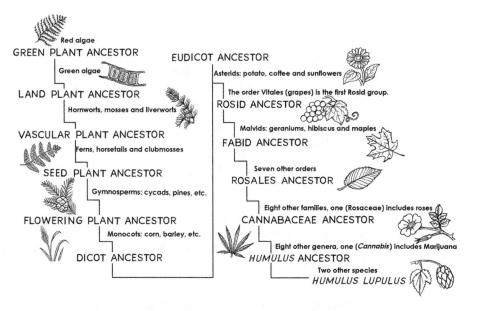

Figure 9.1. Classification of hops (*Humulus lupulus*), showing its position within the various higher categories of plants.

vided into two big groups called the fabids (including hops) and the malvids (including geraniums, hibiscus, and maples). The first group to branch off in the fabids is Vitales, which includes grapes. The eight remaining subgroups (orders) of fabids include Rosales, which includes roses, marijuana, and hops, to name only a few. One of the nine Rosales families is Cannabaceae, home to eight genera. Among these, two share many similarities and are very closely related: *Cannabis* (marijuana), and our old friend *Humulus*. The genus *Humulus* currently has three species, *Humulus scandens* (or *japonicus*), *Humulus yunnanensis,* and the common hop, *Humulus lupulus.*

There are currently about ten genera that belong to the Cannabaceae. Interestingly, *Cannabis* and *Humulus* are each other's closest relative. Mei-Qing Yang and colleagues have used this tree topology to decipher the evolution of important morphological characters of hops and marijuana. Specifically, the leaf arrangements (also known as phyllotaxy) of hops and marijuana are very different from the rest of the Cannabaceae. Most plants have alternate, opposite, basal, or whorled

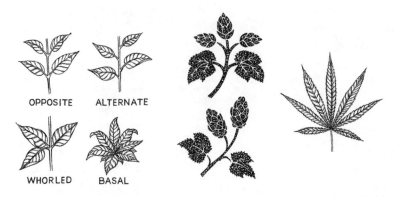

Figure 9.2. Leaf arrangements in plants that are dicots. On the far left are four common arrangements of plant leaves. The two darker images in the middle are hops plant icons that illustrate the opposite and alternate arrangement of leaves. *Cannabis (far right)* has a compound leaf that is palmate.

leaf arrangement patterns (Figure 9.2), and the ancestor of Cannabaceae seems to have had an alternate one. But as the figure shows, both hops and *Cannabis* have a mixed pattern. In *Cannabis* the lower leaves come off the stems opposite one another, while the leaves nearer the top of the plant alternate. In hops, the alternate and opposite arrangements are also both present, although not as hierarchically as in *Cannabis*. This means that both *Cannabis* and hops leaves are palmate in form, made up of projections or lobes that emanate from a single general location at the base of the leaf. The classic nine- or seven-lobed palmate structure of *Cannabis* leaves contrasts with the fuller one-, three-, or five-lobed and webbed palmate leaves of hops (Figure 9.2).

The sexual system of hops is interesting, and of course is key to how these plants are bred. The ancestor of all plants in the Cannabaceae family was probably monoecious, meaning that both the male and female reproductive organs of the plant were found on the same individual. Many of the rest of the genera in the Cannabaceae are also strictly monoecious; but *Cannabis* and hops are among the exceptions in the family. *Cannabis* can regularly be both hermaphroditic (monoecious) and dioecious. In other words, marijuana plants can be male, female, or hermaphroditic, all in a single population. In contrast, hops are mostly dioecious with occasional monoecious interlopers. The reproductive

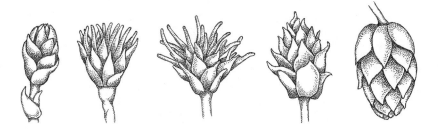

Figure 9.3. Development of the female hop flower, from "pin" stage on the left, to hop cone on the right. Note the spiky projections of the flower in the intermediate stages, and the transformation of those spikes into the tissues of the cone.

habits of these plants are important because only the female plants yield the desired products of marijuana and hops—buds and hop cones, respectively—and only if the plant is a virgin. Marijuana growers have learned to shock or stress their plants to produce feminized pollen for the next generation's breeding and so produce uniformly female populations, and researchers have figured out a way to genetically modify the plant to produce only feminized pollen for breeding. But while hops growers have also tried using stress to do the same thing, it has turned out that pollen from hops plants stressed in this way is not viable. So they just try to limit the males in a population as much as possible.

The male flowers of the hop plant bear the male reproductive structures, namely the stamens that bear pollen. The female flowers include the fruit-producing structures or ova, and it is the small burrs on the female flower that will eventually develop into the hops cone that is such a key ingredient in beer. Ideally, a hops cone should be seedless. In fact, most hops-growing manuals suggest that if a female individual is found with seeds, the male responsible should be hunted down and eliminated. Once the female plant has flowered, the burrs on the flower will develop into the easily identified cones (also known as strobiles) that are illustrated in Figure 9.3.

Hops also have a mixture of perennial and annual behaviors. The plant is perennial in that it can live up to twenty years, but annual in that it reproduces only once a year. The flowers form along vinous structures that botanists call bines. We have so far referred to hops as a vine, but

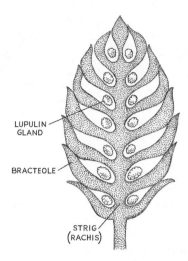

LUPULIN
GLAND

BRACTEOLE

STRIG
(RACHIS)

Figure 9.4. A hops strobile with the outer bracts removed and split, to give a midsection view of the inner structure of the cone. The bracteoles are the greenish sheaths that surround the cone under the bracts. The strig is a part of the stem that extends throughout the cone, and is the point from which the bracteoles emanate. The lupulin glands lie close to the central axis of the cone.

there is in fact a major difference because bines grow helically upward without the assistance of suckers or tendrils to anchor themselves. Instead, many bines have downward-pointing hairs that cling to the structure they are growing around. In hop-growing areas of the world the wooden supports and strings on which the hops grow can be an impressive sight, with bines curling upward to heights of as much as thirty feet.

Hops cones are female flowers that have developed further than the flowering stage. Many plants do this, making fruit to nourish and protect the developing embryo (if there is one). As we saw, growers can trick plants into going into this stage without being fertilized, and hops are no exception. The critical parts of the anatomy of the strobile are shown in Figure 9.4 along with the strig, the lupulin glands, the bracts, and the bracteoles. The bracts are the leafy green sheaths that make up the outer structure of the cone, and their chemical components are not important in beer brewing. The bracteoles are small leaflike protrusions off the strig, which is the stem of the cone. Bracteoles contain oils and resins, as well as tannins and polyphenols, all of which contribute to beer brewing. But the lupulin glands are perhaps the most important part of the cone in this respect. In a freshly picked hops strobile, the lupulin glands will look yellow, and they will feel quite sticky because they are basically blobs of essential oils and resins. These blobs taste bitter, and are the source of the bitterness that hops give to beer.

There are hundreds of different chemicals in a hops cone. By weight, an average hops cone will contain cellulose and lignins (40 percent), proteins (15 percent), total resins (15 percent), water (10 percent), ash (8 percent), tannins (4 percent), lipids and wax (3 percent), monosaccharides (2 percent), pectins (2 percent), and amino acids (0.1 percent). It is not surprising that almost half of the cone is cellulose and lignin, because these are both important structural compounds in plants. They are rather tough molecules (we have difficulty digesting cellulose, even with the help of the bacteria in our guts), and their impact on the taste and odor of the beer we drink is minimal. Of the rest of the compounds listed, the most important are the essential oils and total resins, because they give beer its bitter taste and distinctive aroma. The essential oils also give some beers their fruity, spicy, and floral characteristics. The total resins category includes two major kinds of resin, hard and soft. The soft resins are the part of the total resin that is soluble in the organic chemical hexane. This fraction of the total resins is often specifically quantified for hops strains, because it contains the alpha acids that are important for taste and aroma. The hard resins, insoluble in hexane, are made up of beta acids, slightly different molecules from the alpha acids. The alpha acids concerned are mainly humulone, cohumulone, and adhumulone, while the beta acids are mostly lupulone, colupulone, and adlupulone.

The alpha acids that come directly from the hops plant itself are not bitter. To acquire this quality, they need to go through a chemical process called isomerization. This is achieved by boiling. The structures of the before-and-after molecules are quite different, and the shapes of the small alpha and iso-alpha forms are very important for how humans taste and smell them (see Chapter 11). To cut a long story short, it is the transformation of the humulones into isohumulones that gives beer its bitter taste. Beta acids, in contrast, do not isomerize with boiling. They do so only when they are oxidized, something that brewers strive to avoid because beta-acid bitterness is considered unpleasant.

One of the more important brewing measurements is called the international bitterness unit, or IBU. This is defined as isohumulone in parts per million, and measuring it in beer is a somewhat complicated process. Alpha acids and iso-alpha acids are soluble in hexane and other organic solvents, and if the goal of measuring IBUs is to quantify the amount of isohumulone per unit volume, the solubility of alpha acids in organic liquids can help extract the isohumulone from a measured volume of beer. The beer is measured after the post-wort boiling and hopping steps, at which time the humulone will be in its iso-form.

Here's how it goes. A specific amount of beer is mixed with isooctane. Like many organic compounds, isooctane is not soluble in water, so the water in the solution will separate from the isooctane. Next, to ensure that all the isohumulone is dissolved in the isooctane, the pH of the whole mixture is lowered so it becomes more acidic. This step should dissolve all the isohumulone acids in the beer and isooctane mixture. Again, because isooctane and water do not mix, there will be two phases in the tube in which all these reactions are going on. The isohumulones are in the organic phase, which is easily drawn away from the water phase. A specific quantity of the organic phase is then poured into a small glass or plastic container called a cuvette. This in turn is placed in a spectrometer, a machine that blasts light of different wavelengths through the small amount of solution in the cuvette. The isohumulones in the solution absorb or block the light from reaching a detector lying directly across from the beam of light. A specific wavelength of light (275 nanometers) is used to measure the light absorbance, which is proportional to the concentration of isohumulone in the beer. That measured absorbance number is run through an equation, and voila! We have our IBUs, a measure of the bitterness in the beer.

Most beers will have IBU values between 20 and 60. But this measure isn't all there is to evaluating bitterness: if a beer with an IBU of 60 contains other compounds that mask its bitterness, it can taste less bitter

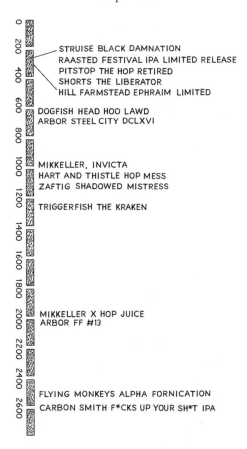

Figure 9.5. International bittering unit values of some beers in the range from 200 to 2600, including the world's bitterest. There are many thousands of beers in the under-200 IBU range. Some of the beers listed may not still be in production.

than a beer with an IBU of 20. Note also that we are referring to bitterness as defined by IBUs, and not to the "hoppiness" that is conveyed by some other qualities of the hops. The international bitterness unit is not a measure of hoppiness, and should not be thought of as such. Most beers in the range of 100 to 200 IBU are already notably bitter, but there are some monsters out there that go all the way up to 2,600 IBU (Figure 9.5). This has led to some controversy about the utility of the IBU measure, since only below the 150 level (perhaps well below) will our sense of taste reliably discern the varying bitterness of modern brews.

Finally, some nuances of IBU and hops require mention. A beer's IBU will diminish if it is stored for too long, which suggests that the iso-humulone tends to degrade over time. In addition, it is now common practice to pelletize hops. This process uses a hammer mill to grind the dried hops into a fine powder, which is then compressed into little pellets that look a lot like animal feed. Which form (pellets or cones) is better for brewing? It remains a matter of opinion. Both whole hops and pelletized hops have their advantages and disadvantages, and in the end, it is six of one, a half-dozen of the other.

So far, we have been treating hops as a single entity. But there are in fact many kinds of hops, suited to different kinds of brewing and to different phases of the brewing process. Some varieties are high in alpha acids and are prized specifically for their bittering qualities. They are typically added early, and include Galena, bred at the University of Idaho just when the craft beer revolution was beginning in the United States, and Nugget, developed a few years later in Washington State. Both of these come in at around 13 percent alpha acids, which is high compared to the 9 percent of an Old World bittering strain like Northern Brewer, bred in the United Kingdom in the 1930s. But while hops in general are renowned for imparting bitterness to beers, most varieties are actually classified in the "aroma" category. Here alpha acids are lower, and more subtle flavor compounds predominate. In the United States aroma hops include the Cascade, renowned for the spicy, floral, and citrusy flavors it imparts; and the Columbia, which is sometimes thought of as an American alternative to the classic English Fuggle. The Fuggle is the backbone of many an excellent English ale, to which it imparts woody, herbal, or sometimes even fruity aromas. Another classic English aroma hop is the Golding, which has a unique flowery spiciness.

Some hops strains have been bred specifically to provide both alpha acids and aroma. Northern Brewer is on occasion placed in this category, along with the American Cluster and the German Perle. Interestingly the Saaz, the classic hop used in the production of the pilsners that are renowned for their clean bitterness, yields only 3 percent alpha acids. Along with such strains as the German Hallertau and Tettnanger, Saaz is often classified in the category of "noble" hops that strongly emphasize aroma over bitterness. That the same hop strain can so easily be placed in more than one category is perhaps less surprising when one realizes just how rampant the cross-breeding of strains has been. For example, while the American Centennial aroma strain is ¾ Brewers Gold, it is also ³⁄₃₂ Fuggle, ⅙ East Kent Golding, ¹⁄₃₂ Bavarian — and ¹⁄₁₆ Unknown, even though the strain was bred recently, in the 1970s.

The variety of hops strains on offer is thus nothing short of bewildering, which is probably one of the reasons that biologists have made less progress than they have in the cases of yeasts and barley in approaching the "mother of all hops" problem. Still, a few studies have brought both molecular and other techniques to bear on the issue of hops relationships. One very interesting nongenetic approach was taken by Michael Dresel, Christian Vogt, Andreas Dunkel, and Thomas Hofmann, who looked at 117 chemical characteristics of around ninety hops strains. They used high-performance liquid chromatography (HPLC) to obtain chemical information on the more than one hundred chemical compounds known to be in hops. In HPLC a solution is run through a column that separates the various chemicals in it. These can then be characterized using the spectrophotometry already described. Dresel and his colleagues used an HPLC approach designed to best separate the compounds for each strain and to deliver quantitative data on them, then used these data to construct a pedigree of the ninety or so hops strains, of which ten are shown in Figure 9.6.

Atsushi Murakami and colleagues have used several kinds of DNA sequence analysis, including DNA sequencing of targeted genes in the chloroplast, to examine hops plants from more than forty sites across the globe. Their findings suggest that there are two major lineages within the species *Humulus lupulus:* a Eurasian group of strains, and an Asian/New World (North American) group. Yet even here, the boundary between the two groups is fuzzy, because the Chinese samples did not fit neatly into the analytic categories. And that is about as much progress as has been made at this time.

Researchers have demonstrated that there is enough variation in

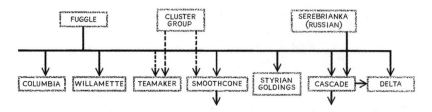

Figure 9.6. A small portion of the hops pedigree by Dresel et al. (2016), showing the relatedness of only one-tenth of the strains used in the study. The arrows pointing downward represent connections to the rest of the hops pedigree.

the genomes of the various hops strains that, in the future, DNA fingerprints might be used to identify the origin of otherwise unidentified hops samples. But genomic analysis of hops plants is still in its infancy because the first draft genome of a hops plant was generated only in 2015, and first drafts are notoriously incomplete. Even so, the availability of this genome should open doors for future genetic and genomic research, especially because a hops genome database already exists (HopBase). This database will eventually be used to explore such aspects of hops as yield and the genetics of its resistance to fungal and viral infections—as well as to clarify the biological history of a remarkable plant that is of critical importance to modern brewing.

PART THREE

The Science of
Gemütlichkeit

10

Fermentation

There are "beers" out there that are touted for their bizarrely high alcohol contents, rivaling those of the strongest spirits. But those alcohol bombs are usually made using freeze distillation, which involves artificially removing the less alcoholic portions of the liquid. To preface our discussion of fermentation we preferred to taste something a little more restrained, if not exactly traditional. We chose a 16.9 percent alcohol spice-and-pumpkin ale aged in rum barrels. It poured deep copper, with no head, few bubbles, and a deep viscosity that was not evident until it hit our palates. The alcohol was smooth and rich, just as you taste it in a heavily fruited rum cake. As a control, we tested this amber beauty against a 120-minute hopped ale, at a full 18 percent alcohol by volume. The contrast was complete, but in combination the two beers showed that there are many ways in which clever brewers can turn high quantities of rough barley alcohol to their advantage.

There are many reasons to drink beer, and since one of them is to experience the variety of desirable (and undesirable) effects that its alcoholic content confers, no biological account of beer can avoid at least briefly broaching the amazing chemistry and natural history of the alcohol molecule. If you are not chemically inclined, then you might be satisfied at this point with the following formulation: sugars + yeast = alcohol + carbon dioxide. But in case you would prefer a little more detail, here goes.

Let's start with where the alcohol originates. Since the word alcohol refers to a whole family of organic molecules, many different molecules are technically alcohols. The specific alcohol of interest to beer drinkers, however, is ethanol. As we know from the earlier Milky Way Bar example, molecules of alcohol exist freely elsewhere in the universe, but because free ethanol is rather rare on this planet, to obtain it we human beings have to either find organisms that make it, or laboriously synthesize it in the laboratory. For brewers and wine makers, the organism of choice to turn those sugars into alcohol is the yeast *Saccharomyces cerevisiae,* which happily produces ethanol.

The function of any molecule depends both on the atoms of which it is made, and the way in which those atoms are arranged. The arrangement of atoms in turn influences a molecule's shape (the way it folds), and thus how it behaves. Molecules with the same chemical composition can differ in the way they are arranged in space, and hence can behave differently. Molecules and their atoms—and their electrons—are the stuff of chemical equations.

The first rule with any equation is that its two sides must balance—or there will be some interesting side effects. When writing a chemical equation, we use the symbol of each atom a molecule contains, with a subscript giving the number of times it occurs. For example, carbon dioxide, which has one carbon (C) and two oxygens (O), is written as CO_2. But this shorthand doesn't capture how the atoms of a molecule are arranged. To better describe the molecule's shape, and so help us understand more about its function, chemists like to use "stick and ball diagrams" instead. These diagrams employ round and stick-like symbols that look something like Tinkertoys. Each atom "ball" has a specific number of "sticks" protruding from it, depending on how many bonds that atom can make to adhere to its neighbor atoms. There are several ways in which two atoms can bond together, the most common being the ionic bond formed when two atoms share an electron. Because hydrogen usually makes a single bond, it has only a single stick emerging from it, while oxygen typically has two sticks because it can make two bonds, and carbon has four because it can make four bonds. The number of sticks protruding from a given atom is dictated both by its atomic number, and by the orbits of its electrons; and in stick-and-ball notation carbon dioxide looks like this: $O=C=O$. Note that there are four bonds in total, two each from the oxygen atoms, so that the carbon atom has a total of four.

This notation is flat, but molecules exist in space and have a three-dimensional structure. So we need to discriminate between carbon dioxide's bookkeeping format and its natural (stick-and-ball) form. In this instance, the three-dimensional structure of carbon dioxide resembles its bookkeeping structure, since it is arranged linearly. But many other molecules have atoms that are connected to one another at an angle, giving them a true three-dimensional structure. This is key in brewing, because at the molecular level at which fermentation takes place, the sizes and shapes of molecules are what drive the reactions that produce alcohol. Nature doesn't necessarily care which atoms are involved; rather, it takes its cues from the external form of each molecule.

Like most drugs, alcohol is a tiny molecule. Indeed, with a molecular weight (the sum of its atomic weights) of 46, it is smaller than the smallest prescription drug (hydroxyurea, which has a molecular weight of 76). Alcohols have a central carbon atom, as shown in Figure 10.1. Remember that a single carbon atom can make four bonds, so in the typical alcohol molecule there are four arms emanating from the central carbon. One of these is always the same in all alcohols: an OH (or hydroxyl) group. The R's in the figure may be other organic molecules such as hydrogens, or a more complex molecular side chain called a methyl group (CH_3). If all three of the R's are hydrogens, then you have a methanol molecule (very toxic, and to be avoided because it can cause blindness and death). Simply changing one of the hydrogens in methanol to a CH_3 side group will, however, convert the molecule from the lethal methanol to the usually delightful ethanol.

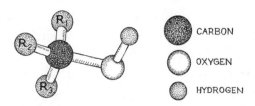

Figure 10.1. *Left:* generic formula for an alcohol molecule. The R's indicate groups attached to the central carbon. The invariant part of the alcohol molecule is the hydroxyl group (OH). In methanol the R's are hydrogens, whereas in ethanol, R_1 and R_3 are hydrogens and R_2 is a methyl group (one carbon atom bonded to three hydrogens).

Two other kinds of alcohol are important to brewers because they are produced when contaminating bacteria and yeasts get into the fermentation process. These alcohols, butanol and propanol, are unwanted because they are toxic to the nervous system. These alcohols are produced by the breakdown of cellulose — something that is hopefully not in a beer brew to begin with.

Yeasts produce the alcohol in beer by breaking down the sugars in the malt. The most familiar of these sugars is the sucrose we use to sweeten our coffee. There are also similar-looking molecules called maltose and lactose. All three are disaccharide sugars, which are made by combining less complex sugars called monosaccharides. (Some even more complex polysaccharide sugars are also of interest to brewers, but are less common.)

The basic structure of a sugar is a ring of carbons. Monosaccharide sugars can have either five (pentose) or six (hexose) carbons in their rings. Two adjacent carbons in a ring will have a single bond between them, and a bond to the carbons next to them, leaving room for two other bonds to emanate from each carbon. In this way, an H or an OH will stick up and down from the carbons in the ring, in a variety of combinations, to balance the chemistry of sugars. Not all sugars will taste the same, because the different groups sticking up and down from the carbons in the ring give each a unique shape that interacts with the taste receptors on our tongues. In Chapter 11, we will delve into how triggering different receptors on the tongue results in taste; but the basic idea is that the shape of the thing tasted (sugar in this case) is the source of the taste.

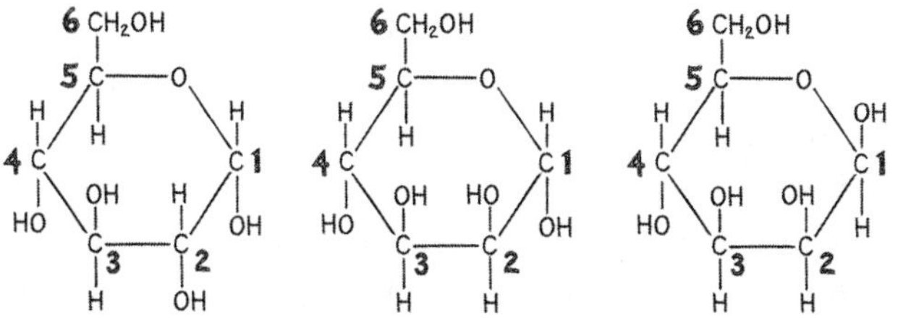

Figure 10.2. *Left:* The chemical structure of glucose. Note that from carbon 1 to 4, the order of OHs is down, down, up, down. Other sugars have different arrangements of the OH groups. The middle drawing describes the structure of unstable mannose (down, up, up, down), and on the right is the structure of bitter mannose (up, up, up, down).

Let's consider glucose, a monosaccharide sugar with six carbons in its ring (Figure 10.2). Chemists number the carbons in the ring as on the face of a clock, but from one to six, starting with the carbon at three o'clock. The groups that come off the carbons can go either up or down, and the order of the H or OH groups on the carbons is critical in defining the overall structure of the sugar. In the glucose molecule, the order of the OH groups from carbon 1 to 4 is down, down, up, down. Flipping the second OH of glucose from up to down makes a down, up, up, down sugar called unstable mannose, which is sweet to the taste but not found in nature. Flipping the OHs on the 1 and 2 carbons to up (giving up, up, up, down) gives us bitter mannose. In this way, two opposite tastes can be generated from the same general structure and chemical makeup, simply by changing the arrangement of the side groups on the sugar ring. There are precisely sixteen ways in which the OH groups on carbons 1, 2, 3, and 4 can be arranged. Each of these will give a unique sugar molecule that, while not differing in basic chemical makeup, can differ drastically in its impact on our taste buds.

Plants have evolved an amazing way of storing the energy produced by photosynthesis. They remove electrons from substances like water and recycle these electrons to make carbon dioxide and other bigger molecules containing carbon, in which energy is stored chemically. Sugars are the end product of this process, so plants can store massive amounts of energy for future use in the form of glucose and other long-chain molecules made from glucose. These longer-chain molecules include starch and cellulose, molecules too big for our taste buds to respond to. These molecules thus lack any taste to us, and cannot be broken down efficiently by our bodies.

Starch is made up of two kinds of molecule. One is amylose, a simple straight-chain molecule in which glycosidic bonds connect one glucose to the next. The second is amylopectin, which, while partly linear, also branches to make larger molecules of starch. Starch is about three parts amylopectin and one part amylose, and once removed from plant cells it appears as a powdery substance. In contrast, cellulose is composed of glucose chains that come together to form sometimes structurally rigid lattices. Paper is made of cellulose, which is also a major component of such foods as lettuce (we are exhorted to include lettuce and other leafy green vegetables as roughage in our diets because the cellulose is barely broken down by our digestive tracts). Significantly, although celluloses and starches are both made of long chains of glucose molecules, they behave quite differently.

These long-chain molecules are the raw materials of the brewer because, quite luckily for all of us, nature has worked out a way to convert them to smaller sugars that can be attacked by yeast to produce alcohol.

When barley grain is harvested, it is chock full of long molecules of starch (intended for the nourishment of the embryo within). These are useless for fermentation, since yeast don't have the enzyme machinery needed to break them down. When the embryo is ready to grow, the grain throws some of its resources into breaking down the long starch molecules into the smaller sugars and smaller starch molecules that the barley embryo can use. The grain itself is equipped with a set of enzymes that it uses to make various kinds of sugars such as glucose, maltose, maltotriose, and other more complex sugars. If the developmental process is stopped early, the enzymes stop working, and the sugars and shorter starch molecules just hang out in the grain.

Maltsters trick the barley embryo into thinking it is ready to grow

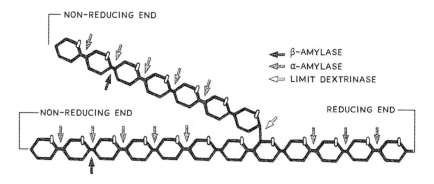

Figure 10.3. How the amylases and limit dextrinase cut starch molecules into single sugar carbon rings. The arrows show where the specific enzymes cut the long-chain starch molecules.

by soaking the grain in water, thereby triggering the enzymatic process that breaks down the long-chain starches for the benefit of the embryo. When the grains are bursting with sugars and short starches, the maltster stops the process by heating and drying the grain, which is then toasted in an oven. The drying time can be varied to give the malted grain the desired color and taste. By adjusting exactly when and how these steps are accomplished, the maltster can control the ratio of starch to enzyme, which is important for the next step of the brewing process.

Mashing helps get all the sugars out of the sprouted grain (barley and anything else the brewer wants to use). There are several steps to mashing, the first of which involves soaking the malted barley in cold water. This procedure will cause gelatinization, which is not essential for releasing sugars, but does speed up the process when the mash is then heated and the grain swells and gelatinizes in earnest. Mashing, which is performed at a range of temperatures, activates the enzymes in the malt that will convert the long-chain starches to the small starches and sugars that the yeast can use. These enzymes (alpha amylase, beta amylase, and limit dextrinase) act like little machines that slide along the long-chain starches, snipping the bonds between the sugar rings. Figure 10.3 depicts these three enzymes at work on a starch. Starch can be strictly linear (amylose) or can branch (amylopectin). The two enzymes that snip those starch molecules are the amylases, which work

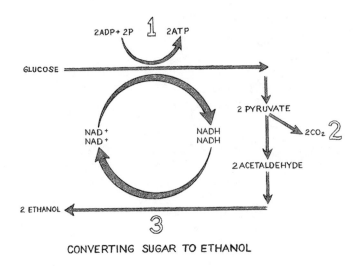

CONVERTING SUGAR TO ETHANOL

Figure 10.4. The fermentation process. The three numbers (1, 2, 3) indicate the two "machines" (1 and 2) and the chemical reaction (3) involved in converting sugar to ethanol in yeast.

on both amylose and amylopectin. The third enzyme, limit dextrinase, cuts at the branching points in amylopectin, and reduces the size of these starch molecules by getting rid of the side chains. The resulting soupy solution is packed with single-ring sugars such as glucose, and is known as wort.

This mixture of liquefied grain and sugars is then made available to the yeast (usually after hopping is done). The yeasts that we discussed in Chapter 8 have evolved to take in the small sugars as food, and to break them down using another suite of enzymes. Figure 10.4 shows the three subprocesses (1, 2, and 3) that brewing yeasts use to convert sugar into alcohol. The conversion is actually carried out by two complex molecular machines (1 and 2) and one simple chemical reaction (3).

The first machine makes a small molecule called pyruvate from larger sugars such as glucose (Figure 10.5). Pyruvate is converted into a smaller molecule, acetaldehyde, by the second enzyme machine. And finally, acetaldehyde is converted into alcohol via a simple chemical reaction. The first machine is a complicated one, involving nine proteins linked together into a larger machine that carries out a process called

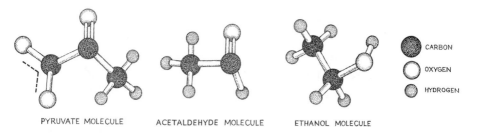

PYRUVATE MOLECULE   ACETALDEHYDE MOLECULE   ETHANOL MOLECULE

CARBON
OXYGEN
HYDROGEN

Figure 10.5. The three products of fermentation. The dotted line on the left in the stick-and-ball figure of pyruvate indicates that the two oxygens that bind to the carbon at the vertex share an electron. This arrangement makes pyruvate very reactive. In making alcohol, the enzymatic machine that breaks down pyruvate is called a decarboxylase, because it removes a carboxyl group and emits carbon dioxide (CO_2). The emission of CO_2 produces bubbles, or carbonation. The acetaldehyde in the middle looks a lot like alcohol, except for the double bonded oxygen bound to the right carbon. To get to the final structure of ethanol, *right,* a hydrogen molecule needs to be added to break the double bond. To become ethanol, all an aldehyde needs to do is acquire one proton, which comes from the classic proton donor molecule NADPH.

glycolysis. The functions of the nine enzymes that are part of the first machine are mostly to add a molecule such as phosphate (P) to the reacting molecule, or to break a bond. An important molecule in all of this is called nicotinamide adenine dinucleotide phosphate oxidase (NADPH), which, with the help of adenosine triphosphate (ATP), helps move protons around molecules.

So far, we have seen how yeast fermentation works. But yeasts are not the only players. Some bacteria have also learned the trick of fermentation. Like yeast, bacteria create pyruvate molecules by glycolysis, but they have their own method of dealing with the pyruvate. In the absence of oxygen, or of the enzyme aldehyde decarboxylase (something yeasts have but bacteria do not), the reactive pyruvate will grab an electron from NADPH, to produce NADP. This added electron reduces the pyruvate and, as represented in the diagram, changes it into the small molecule known as lactic acid. Note that the change occurs in the middle carbon of the pyruvate molecule. What has happened is that the doubly bound oxygen has taken up hydrogen (has become reduced, as chemists say) to form an OH group that sticks out of the middle carbon. This process produces NADP, which can be recycled via glycolysis. In this way, bacterial cells have found an economical and evolutionary distinctive way to deal with their electrons.

Lactic acid and ethanol are products of bacterial and yeast fermentation, respectively; but although they possess fairly similar chemical makeups, these two molecules taste very different due to their distinctive molecular shapes. Bacterial fermentation is usually considered a defect in beers, but not always. In fact, some traditional beers such as the German Berliner Weisse are produced by adding *Brettanomyces* ale yeast (a yeast that produces a wide array of tastes) and bacterial species of the genera *Lactobacillus* or *Pediococcus*). *Brettanomyces* can enliven the taste of beer in distinct sensory ways. In addition to alcohol, *Brett* produces three major chemicals during fermentation. These three compounds—4-ethylphenol (antiseptic odor/taste), 4-ethylguaiacol (smoky odor/taste), and isovaleric acid (cheesy odor/taste)—are responsible for the distinctive characteristics of *Brett*-brewed beers. *Brett* ferments much more slowly than typical brewers' yeast, so fermentation times are much longer. Naturally enough, manipulating these various natural fermenters is one of the brewer's go-to tricks. By using different yeast strains, various malting and mashing techniques, and by adding or allowing other organisms that use sugar, brewers can create an amazing range of beers with a diversity of characters, tastes, and aromas.

But how much alcohol is in each variety? In Chapter 6, we reported on the technique of measuring the specific gravity of a brew both before fermentation and after, in order to quantify alcohol content. By assuming that the sugars in the wort stage of the brew are converted only to alcohol and carbon dioxide, the after specific gravity should reflect the conversion of sugar to alcohol per unit volume, and hence give us an estimate of how alcoholic a particular brew is. A simple equation converts these specific gravity measurements to alcohol by volume (ABV) and alcohol by weight (ABW). ABV is the number you will most often see quoted. As the alcohol content goes up beyond 9 percent, there is greater and greater discrepancy using these approaches; but up to 9 percent alcohol the estimates usually agree and are pretty accurate.

Here is the equation for alcohol by volume: $132.715 (OG - FG)$, where OG is the starting specific gravity, FG is the final specific gravity, and 132.715 is the "magic number" or a constant that appears to do the trick for converting between specific gravity and percent alcohol. So, if your initial OG is 1.066 and the final gravity is 1.010, then the ABV will be 0.056×132.715, or 7.43 percent ABV.

ABW is calculated using the same input numbers but a different "magic number" that reflects the alcohol percentage by weight. The ABV can be converted to ABW with this simple equation: $ABW = ABV \times 0.79336$. So, for the same brew, we get ABW equal to $7.43 \times 0.79336 = 5.894$ percent.

Although these calculations are important to brewers and beer drinkers everywhere, the main factor for the ordinary beer fan to keep in mind is that what you are drinking is the product of billions and billions of chemical reactions, all managed by living creatures. Your beer is a living, breathing creature.

11

Beer and the Senses

The squat brown bottle and the neo-Gothic type on the label promised contents that were archaic and unusual, as was indeed the case. This was a classic "Smoke Beer" from the central German region of Franken, brewed with lager yeast but using a dark malt that, in immemorial fashion, had been heavily smoked over a beechwood fire. It poured a deep, dark chestnut color, but was shiningly clear, like a peal of Franconian church bells. The head dissipated quickly, but it left stringy traces on the side of the glass that reminded us of wisps of rising smoke. On the nose and palate the beer overwhelmed, with pungent smoky aromas and flavors that lingered until the last drop had left the glass. We could close our eyes and almost hear the wood fire crackling in the background. This beer might have been brewed following the purity laws, but it was about as far from a modern pilsner-style lager as you could get.

The bombardment of your senses begins when you pull the beer bottle out of the cooler, feeling the coldness of the bottle and seeing the color of the label. But your senses have only just begun their journey with that beer. Both your sense of vision and a possibly less appreciated sense of temperature detection are already whirring away, sending messages to your brain for interpretation: Is this beer the one you wanted? Is it too cold? When you finally uncap the bottle, several sensory events occur. If you do it right, you can get a nice pop, followed by a hiss caused by the release of the carbon dioxide that had pressurized the beer in the bottle. As you pour your beer into a glass, your vision and hearing kick in again, taking in the color of the brew, its hue and transparency (or lack thereof), and the gurgle as the beer fills the glass. Next, your nose is inundated with aromas as you raise the beer to your mouth. As your lips touch the glass, another rush of neuronal information tells your brain that something cold is coming down the hatch. Touch-receptor cells in your lips then guide the glass to its proper resting place, and when you tip it, things get serious. The taste-receptor molecules in the taste buds of your tongue start to gather up the molecules that wash across them, inundating your brain with information about the saltiness, sweetness, bitterness, and sourness of the beer (and, if you are lucky, about its "umaminess"—umami is a fifth, "savory," taste category for which we have receptors). You will also taste the carbonation in the beer because there are receptors for this sensation, too; and you might even be able to taste the alcohol if it is in sufficiently high concentration.

When the beer hits the back of your throat it will start a new phase of its journey, although along the way your cold receptors will kick in again, and some of the taste receptors in the back of your mouth will also be stimulated. As you swallow, a backwash of tastes will bathe your tongue and send even more information to your brain. If it is pleasant-tasting beer, your brain will start to like it, and you will raise your glass again. If it tastes terrible—if, for example, it has skunked or gone flat— you will most likely shun it. Either way, it is safe to say that in the act of swallowing your beer your brain has been inundated with information from virtually all your basic senses. It will only be later, after the beer has been absorbed in your gut, that your brain will start to be influenced by its alcohol content (see Chapters 12 and 13). Meanwhile, your brain has been plenty busy.

All the information the brain receives from the outside world comes from the sensory systems, via the electrical impulses that are the "currency" of the nervous system. Those electrical impulses are part of a very effective physiological system that transfers information from far-flung parts of the body to the brain, and back. The eminent scientist Francis Crick once said that the brain's output (including our own peculiar form of consciousness) is "entirely due to the behavior of nerve cells, glial cells, and the atoms, ions, and molecules that make them up and influence them." This statement holds equally for drinking a beer. Our response to beer is nothing more, and nothing less, than electrical impulses coursing to specific parts of our brains, where what is happening is interpreted in amazing perceptive detail.

We can look at beer as a source of signals that our sensory systems capture and interpret via perception. Vision, taste, and smell—three of the "big five" senses—are obviously involved in enjoying beer; but hearing, touch, and temperature perception are also important in our sensory experience of the beverage. In Chapter 13 we will explain that alcoholic beverages can also significantly affect our balance, so that drinking beer will ultimately affect the full complement of our senses.

The pop of the opening bottle, and the swoosh of the beer can, are nothing more than waves of disrupted air that are collected by the outer ear, which acts like a natural funnel to collect the sound for its journey into your inner ear. There are about a dozen ways in which sound can be measured; here we consider frequency (pitch) and volume (loudness). Like all waves, the displaced air caused by the pop of the bottle will have both a frequency and an intensity. The wave frequency is measured in units called hertz (Hz), and the intensity is measured in units called decibels. The human range of detectable frequencies runs from around 20 to 20,000 Hz. At the low end, the low notes of a pipe organ are about 20 Hz, while normal human speech is at around 500 Hz. Mariah

Carey's whistle singing at the end of *Emotions* is about 3,100 Hz, and the frequency of sound from cymbals clashing comes in at about 10,000 Hz.

The sounds generated by opening beer bottles are relatively high-pitched, in the thousands of hertz. The unit of volume for sound (the decibel) is a relative measure, because the distance from the source of the sound is critical. Yet it is probably more relevant than Hz for opening a beer bottle or can. The typical range of decibels that humans can tolerate is from 0 to about 140 or so, when the loudness is so extreme it can physically endanger the structure of the inner ear. Whispering happens at around 20 decibels, normal face-to-face conversation is carried on at about 60 decibels, jackhammering emits sound at about 100 decibels, and a jet takes off at about 130 decibels. We figure that the sound of opening a beer bottle has a loudness of about 50 or 60 decibels. Pouring a beer has lower hertz and decibel levels than opening a bottle; but it is still audible, and we would think it odd if we saw a bottle opened nearby without hearing it too.

The sound waves from the opening bottle, the glug of the carbonation during the pour, and the delicate fizz as the beer sits in front of us all reach the inner ear and are collected by the membrane of the eardrum. The vibrations of the eardrum then mechanically interact with three tiny bones in the inner ear, commonly known as the hammer, the anvil, and the stirrup. Through a chain of mechanical interactions from the eardrum, to the hammer, to the anvil, to the stirrup, the characteristics of the wave hitting the eardrum are further transmitted to a structure in the inner ear called the cochlea (Figure 11.1). The cochlea is lined with hairs that are embedded in neural cells and is filled with fluid that reacts as the stirrup moves, like a piston, to mechanically transfer the characteristics of the sound waves to the cochlea. As the fluid in the cochlea moves, the hairs respond by bending in specific ways for specific sounds. The neural cells that the hairs are connected to respond accordingly, and the information is sent to the brain via the electrical impulses we discussed earlier.

How we interpret the sounds is a matter of memory and emotion, and of some tricky characteristics of our brains that have been studied by the social scientist Charles Spence. He and his colleagues have looked at everything from how the names of chocolate bars cor-

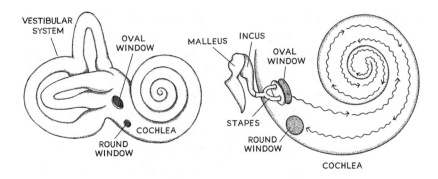

Figure 11.1. *Left:* The cochlea and its relationship to the vestibular or balance system (semicircular canals). *Right:* The relationship of the three inner-ear bones malleus (hammer), incus (anvil), and stapes (stirrup) to the cochlea and to the round and oval windows.

relate to perceptions of their taste, to how consumer preferences are affected by the color of soda cans or by how noisy food packaging is when it is opened. They examined a plethora of bottle-opening and pouring sounds, including the opening pop (the Grolsch ceramic cap was a favorite example). Spence and colleagues divided these into three subsounds: temperature, carbonation, and viscosity. Believe it or not, the sound of a pour can tell a knowledgeable drinker if the beer is cold or warm, or even well carbonated; where viscosity differences are large enough, these can also be detected audibly. Just by listening to it pour, you will be able to form a pretty good impression of the beer you are about to drink.

The pour thus primes us for our first swallow of the brew, as does its appearance in the glass once poured. Let's say you are staring at a newly poured glass of lager. Light of all wavelengths will be hitting the lager from many directions, some of it bouncing off and some being absorbed. The lager in the glass is a golden yellow, so that light that is not of a wavelength between 570 nanometers and 590 nanometers (the wavelength of yellow gold) will be absorbed by the beer. Any light between 570 nm and 590 nm will be reflected, and this is the light that will collide with the retinas of your eyes, letting you know the color. Information about light reflecting from objects around the glass, about light traveling through the glass (if you are looking at a clear lager), and about shadows around the glass, will also be sent on from the retina to the brain. This is information about the shapes and objects that are in your field of vision.

Our eyes are complex structures, of which the retina is the most important here. The retina is a field of cells at the back of the eye that some scientists contend is actually part of the brain. It contains two major kinds of cells, known as rods and cones: there are some 120 million rods and around six to seven million cones. These cells transmit information about light coming in from the outer world to the brain, via the optic nerves. The rods collect and transmit information about the general characteristics of the light being viewed. They function better in low light than the other cells of the retina, and hence provide night vision. Cone cells come in three major forms—red, green, and blue—named for the colors that they are responsible for detecting. Both rod and cone cells collect information from light via molecules called opsins. The major kind of opsin in rod cells is called rhodopsin. The red, green, and blue cone cells have their own kind of opsins called red opsin (L cone), green opsin (M cone) and blue opsin (S cone).

The opsins are proteins embedded in the cell membranes. Each one has a small molecule called retinal (a vitamin A relative) sitting in a pocket of the protein. Retinal is what researchers call a chromophore, a molecule that reacts when hit by light. The retinal response causes a chain reaction in the rod or cone cell, producing an action potential that travels to the brain. Each opsin reacts to an optimal wavelength of light, so that information from light of different wavelengths makes its way to the brain via this system.

Let's put this sensory experience in the context of our glass of lager. The light of wavelength 570 to 590 nanometers hits the retina, exciting cone cells. Pure green light would set the green opsins humming, and signal that the object is green. But there is no opsin that has an optimal wavelength of 570 to 590 nanometers. Instead, both red and green opsin cone cells start to react—but at lower levels than they would if they were being hit with pure red or pure green light. The brain will thus interpret the beer's color as yellow-gold. And now you are ready to bring the beer to your lips.

One of the most iconic drawings in all of neuroscience is that of the "homunculus" (Figure 11.2). This image derives from the work of a neurosurgeon named Wilder Penfield. While his patients' brains were open to manipulation on the operating table, he would "tickle" certain brain parts, then either ask them what they felt or observe a certain part of the body twitch. But weren't the patients anesthetized? Well, no: because the surface of the brain itself has no pain receptors, brain surgery can be done without general anesthetic. Simply watch the last fifteen minutes of the movie *Hannibal,* and you will get the gist. This characteristic allowed Penfield to map the parts of the brain responsible for sensory and motor function, and to represent them on the homunculus proportionately to their importance. For beer drinkers, the important point is that the lips and tongue of the homunculus are way out of proportion to their real size on our bodies, and this enlarged amount of neural real estate is relevant to how you sense that glass of beer you are raising to your mouth.

Like hearing, touch is a mechanical sense, and we have several kinds of highly specialized cells devoted to detecting objects with which we come into contact (Figure 11.3). The main ones are called Meissner's corpuscles, Merkel cell-neurite complexes, Ruffini endings, and Pacinian corpuscles, and they are embedded in the bottom layer, or dermis, of the skin. The critical ones for the glass of beer hitting our lips are the Meissner's corpuscles, because these cells detect light touch. And it turns out that our lips are packed with receptor cells of this kind.

Meissner's corpuscles are very sensitive indeed. If one is distorted by force (such as by lightly touching a glass of beer to the lips), it will produce an action potential that will travel to the brain and allow the brain to map what is touching where. Meissner's corpuscles are also

Figure 11.2. The sensory homunculus, showing the amount of neural real estate dedicated to sensing in different parts of the body. The enlarged parts of the body indicate a larger area of the brain dedicated to sensing. For instance, the lips are enlarged relative to the nose, and hence a larger part of the brain is dedicated to sensing with the lips than the nose.

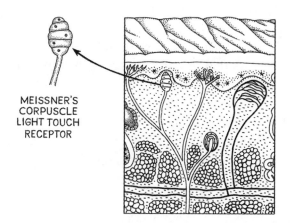

MEISSNER'S
CORPUSCLE
LIGHT TOUCH
RECEPTOR

Figure 11.3. A Meissner's corpuscle, and the top layer of skin in which the receptor is embedded.

found in large numbers in fingertips, where they facilitate the manipulation of objects such as beer glasses. As the glass hits the lips, and the Meissner's corpuscles relay information to the brain about the position of the glass, you can accurately and efficiently deliver the brew to your mouth. Before the beer gets poured down, however, you probably will realize that there is a nice smell emanating from the glass, bringing us to the sense of olfaction.

Humans are often said to have a poor sense of smell relative to other animals, especially those (like dogs) that do most of their communicating with their environment through smell. But while it was accordingly long believed that humans could smell on the order of only ten thousand different kinds of odorants, recent work by Andreas Keller and his colleagues has indicated that we might be able to detect up to a trillion unique smells. Olfaction is what scientists call a chemoreceptive sense, because—unlike hearing and sight, which detect waves, or touch, which detects a mechanical distortion of sense cells—olfaction (and taste) respond to molecules and chemicals that are floating in the air, or are present in the liquids and solid foods we ingest. The molecules and chemicals that we perceive as odors have specific shapes: the flowery smell in hops, for instance, is caused by a small molecule called linalool, whereas the woody odor of a hoppy beer is caused by a molecule called beta-ionone.

Molecules that produce different odors will have correspondingly different structures, and generally, the more similar the molecules, the more similar the odors we perceive. Odor is, then, based on the capacity for the cells in our nasal passage to recognize these shapes, and to transmit information about them to the brain. The roof of the nasal cavity is lined with olfactory reception cells. These receptor cells are connected via nerves to the olfactory bulb of the brain. The olfactory cells themselves have proteins embedded in their membranes that loop in and out of the membrane seven times. One end of the protein has a specific structure that will recognize and interact with the odorant molecule.

There is some controversy as to whether the odorant molecule physically reacts with the receptor protein, or if some other physical phenomenon such as vibration is responsible for the interaction. Either way, the interaction of the odorant with the receptor protein causes a cascade of chemical reactions in the odorant receptor cells, initiating an action potential that then travels to the brain for interpretation. This set of interactions means that we need to have a large repertoire of proteins that can be embedded in the membranes of olfactory receptor cells. Humans have about four hundred olfactory receptor proteins, compared to elephants, which have nearly two thousand, and dogs, which have eight hundred.

Have you ever wondered why beer has such a pleasant smell in the first place? After all, some foods that we eat are not notably pleasant to the nose. But beer seems to have combined a wonderful product with an innately attractive smell. Joaquin Christiaens and his colleagues have shown that the key lies in the yeast. That sweet smell emanating from our glass of beer as it nears our lips comes from two small molecules called ethyl acetate and isoamyl acetate. Both are made by yeasts, as Christiaens and colleagues showed by making a yeast strain that lacked an enzyme critical for their biosynthesis (Figure 11.4). The emission of these two odorants by yeasts is clearly no accident, and it is quite probable that these tiny organisms produce them to entice the fruit flies with which they have been co-evolving for hundreds of millions of years—and that help the yeast to disperse. We seem serendipitously to like the smell of beer—but it is the fruit flies the yeast are trying to attract!

The Master Brewers Association of the Americas recommends that its members use a device called the "flavor wheel" when assessing the taste of their brews. This ingenious contrivance was created in the 1970s by Morten Meilgaard of the American Society of Brewing Chemists, and it has gone through many iterations since then. The flavor wheel attempts to tease out the major flavors of beer to help in comparing different products. Radiating from the center of a circle, it identifies major

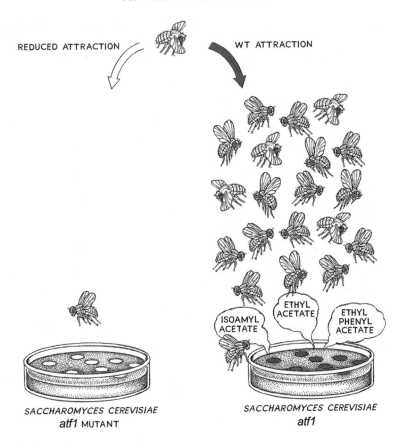

Figure 11.4. The experiment by Christiaens and colleagues to show the attraction of fruit flies to ethyl acetate and isoamyl acetate. The mutant atf1 strain is missing an enzyme involved in synthesizing ethyl acetate and isoamyl acetate. When fruit flies are exposed to the mutant yeasts, they shun them; they are instead attracted to the normal ("WT" or "wild type") yeasts that can synthesize ethyl acetate and isoamyl acetate.

taste classes (aromatic, caramelized, fatty, oxidized, and so forth), and subdivides each one into finer categories (such as grapefruit, caramel, farmyard, funky, burnt tire, and baby sick/diapers, the last of which we hope never to encounter in a beer). In this way, a beer identified as having a "cereal" flavor might be "grainy," or "malty," or "worty."

But while the flavor wheel certainly provides an excellent starting point for discussing the flavor profile of any beer, the very proliferation of different versions of the device underlines just how subjective the problem is that it addresses. Quite simply, many beers will taste very different to different people, simply because humans vary so greatly in their detection of taste. Which means that we should probably declare our own preferences. Throughout this book, we have applied adjectives to various kinds of beer. Adjectives inevitably imply judgment, and for the record we confess we like beers with distinctive and even assertive characteristics. We don't mind if the malt or the hops predominate, but we do find that a "spiky" form of the flavor wheel is preferable to the relative blandness of a beer in which specific flavors do not stand out. We repeat that this preference is completely subjective, and that your own partiality may be entirely otherwise.

Taste receptor cells are found in bundles of anywhere from thirty to a hundred cells, known as taste buds. Like olfaction, taste is a chemo-receptive sense that uses the lock and key mechanism we discussed for olfaction to recognize different tasting molecules; and like olfactory receptors, taste receptors lace their way through the membranes of taste cells seven times, and react to the small taste molecules coming in from the liquids we drink and the foods we eat. Five major categories of taste receptors transmit the tastes of sweet, sour, bitter, salty, and umami (the savory taste of glutamates, like the monosodium glutamate often sprinkled throughout Asian food). And what we perceive as taste is a mixture of the responses of these different kinds of receptors to any given food. The taste buds are, in turn, bundled into structures called papillae, which are physically visible if you look at your tongue very closely in a mirror. The papillae are mainly located toward the front half of the tongue and are sparser at the back.

Human populations harbor three basic kinds of tasters—hypo-tasters, tasters, and super-tasters—roughly in the ratio of 1:2:1. There is also a very rare category of super-supertasters. The number of taste cells on the tongue dictates whether someone can taste, super-taste, or not taste much. Telling whether you are a supertaster, taster, or hypo-taster is a simple matter of counting. Here's how you do it. Take a piece of three-hole-punched paper, and cut out a small square containing one of the holes. Put some grape jelly in your mouth, or take a sip of dark red wine or grape soda, and make sure your tongue gets bathed in the

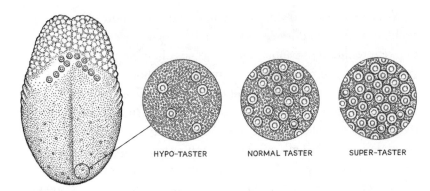

Figure 11.5. *Left:* The areas of the tongue where the taste buds and papillae reside. The circle indicates where the hole from the paper is placed in testing for taste-bud density. Hypo-taster, taster, and super-taster tongues are shown (from left to right) in the three diagrams on the right.

purple stuff. Place the punched hole over your tongue, anywhere near the tip, and look in the mirror. You should see a bunch of little purple mushroom-like structures that you can count. If you can see fewer than fifteen papillae, you are more than likely a hypo-taster, whereas from fifteen to thirty papillae would suggest that you are a taster, and over thirty papillae would indicate that you are a supertaster or even a super-supertaster (Figure 11.5).

Craft beer makers brew some wildly hoppy beers, all of which we enjoy immensely, making us normal tasters. Supertasters will most of the time avoid drinking hoppy beer because they find it incredibly bitter. They certainly tend to steer away from drinking hoppy beers like IPAs, and even lagers are sometimes irritating to their taste buds. Both of us are also immune to the burning sensation on the taste buds that can be caused by the alcohol in a strong beer, whereas a supertaster will report a burning feeling when his or her lips touch a high-alcohol beer, and will hardly ever enjoy a hard liquor. Hypo-tasters, by contrast, will easily tolerate extreme bitterness, but will not be able to tell the difference between a Columbia-hopped beer and a Cascade-hopped beer—which is something that a supertaster will easily do, although he or she will typically find both unpleasantly bitter.

Because of these inherited differences in tasting ability, it is the normal tasters who have all the fun with hoppy beer. But this doesn't mean that supertasters and hypo-tasters can't condition themselves to enjoy alcoholic beverages. They can; and supertasters can even employ their ability to advantage: it has been suggested that most upscale chefs are supertasters who have learned to apply their supertasting abilities to the creation of novel dishes. Even normal tasters, though, usually prefer their beers to be reasonably well balanced among their taste receptors. So while sour beers have recently become very popular, anyone who has tasted a really sour farmhouse ale will recognize that, even as they enjoy it, his or her sour taste receptors are going crazy relative to their bitter and sweet receptors. Still, it is difficult to discourage creative brewers from experimentation—consider the current craze among some for brewing beers with sea salt. We wonder if umami beers are on the way.

As organisms on this planet go, human beings have fairly limited senses. We see in only a narrow range of light wavelengths; we have a limited field of vision; we see in stereo only in a small swath of our field of vision; and we have some very quirky anomalies when it comes to detecting colors. And that is just for vision, the sense that is supposed to dominate in our sight-oriented species. With respect to taste and smell we similarly have reasonable acuity, yet are surpassed in both by many other mammals. These limits point to a major lesson of evolution, which is that natural selection does not strive for perfection, but instead goes for practical solutions. The solutions that our species has settled for make us far from optimized for anything, but they are wholly adequate for our daily perception of the outside world. And quite coincidentally, they have made us excellently qualified to appreciate beer, which hits most of our sensory systems spot-on.

12

Beer Bellies

The ultra-light beer flowed smoothly into the glass, whereas the imperial stout practically had to be spooned out of the bottle. This difference was hardly a surprise, since the ultra promised a mere 96 calories and the stout weighed in at a whopping 306. For a dedicated dieter there was no contest here, and there wasn't one for a pair of dedicated beer lovers, either. The ultra-light came across as recognizably a beer, but little more. The stout, by contrast, blew us away with its density, complexity, and long, lingering finish. Was the difference worth the calories? We knew what we thought.

Beer is delectable, and in moderate quantities it has entirely satisfying effects on the brain. But the chemicals and molecules that the brew brings into our bodies must also, alas, be metabolized. And the sad news is that most of the chemicals involved don't belong in our bodies in the concentrations that beer delivers. At best, they tax the human metabolic system to its limits. In Chapter 13 we consider the effects that beer has on the brain; in this chapter we report on the impact beer has on the rest of the body.

If the purpose of the fermentation process is to produce alcohol, the bottle in front of you will probably have a decent percentage of alcohol in it. Remember too that the other product of fermentation is carbon dioxide, so your beer will also contain some of the gas molecules that give the beverage its fizzy taste. If the yeast cells in the brew have worked properly they will settle as sediment on the bottom of the fermentation tank and will be a major source of the molecules floating around in the fluid. In most cases the brewer will either filter out the yeast layer, or heat-kill it by pasteurization. Typically, though, home and craft brewers do not filter or pasteurize, but rather decant their beers away from the yeast, leaving some of the viable (and highly nutritious) yeast in the brew. During fermentation some yeast will have died of natural causes, and their broken-down cellular components will also be floating around in the beer. The molecules in the cellular remains of yeast are varied, and include cell membranes (lipids), DNA, and the long-chain carbohydrates that kept the yeast alive. All these components of the beer we ingest are used by the body—for better or worse.

Mammals have evolved an efficient but convoluted way of digesting the food and drink they take in, and because the human digestive system is the product of evolution, it has some very quirky moving parts. These quirks occur because natural selection does not strive for either the perfect design, or the optimal outcome. Rather, as we have already noted relative to the senses, evolution is a process that simply finds solutions. Other aspects of it also contribute to the complexities we see in nature. For a start, variation is needed for natural selection to work, and organisms cannot simply conjure up new and useful variations to solve the problems of survival. Populations of organisms are stuck with the variation they have naturally acquired by random processes of mutation in the genes. Moreover, evolution is not always directional. Although it was commonly accepted in the 1940s and 1950s that natural selection inched gradually toward ever more favorable states, since the 1970s it has been recognized that chance events have had an equal, if not greater, impact on evolutionary histories. As a result of these evolutionary vagaries, our body systems are far from optimized in any engineering sense.

Our digestive system extracts nutrient molecules from the food we consume, allowing us to acquire the energy we need to maintain our basic metabolic and bodily processes, to move around, and, very importantly, to feed our energy-hungry brains. Our digestive system also distributes, to other organs of the body, molecules that do not contribute to energy generation. Still, when we talk about digestion we are usually thinking about energy production, which is where calories enter the discussion. Those calories are a slippery notion. They aren't something we can touch or feel, like fat or ethanol. In fact, although calories can be a good measuring stick for how we metabolize our food and drink, and burn energy by being active, they are entirely conceptual: a calorie is defined as a unit of heat or energy, specifically the amount of heat required to raise a single gram of water one degree centigrade. We should note that the calories we see listed on food packaging should actually be multiplied by a thousand—in other words, they are kilocalories—although we will continue to refer to them in the familiar way.

Things other than food can have calories. For instance, the gas your car uses has a specific number of calories, depending on how much is in the tank. The energy or heat in the measured calories comes from "burning," or metabolizing, the source. Different sources of calories are burned in different ways. The materials that our bodies burn to acquire energy from beer consist of ethanol, protein, and carbohydrates. In food more broadly, there are many other sources of energy, with fat probably being the most important. Each source of energy has a specific quantity of calories to contribute to our energy reserves. For each gram of fat consumed, our digestive systems deliver to our bodies some nine calories; for each gram of protein or carbohydrate, it delivers four. Ethanol is converted into energy at seven calories per gram. The calories from pure ethanol are sometimes considered to be nutritionally empty, because they are not delivered to the body with nutrients.

When you are walking, running, or using your brain, your body is using molecules it has ingested to generate energy (measured in calories). If these molecules are not freely available at the time, your body needs to get them from storage, or it will cease to function. That storage is in the form of fat, into which all energetic molecules not immediately needed are converted. This way, the energy from food we ingest is pretty efficiently stored. One might think that, since ingested fat has the highest caloric content (nine calories per gram of intake), most of our problems with overweight or obesity would be caused by fatty food. If this were the case, we could drink relatively fat-free beer to our heart's content, with minimal worry about becoming overweight. But we would, alas, be dead wrong, because the breakdown of fatty foods proceeds very differently from the breakdown of ethanol and carbohydrates, the source of the calories in beer.

The carbohydrates (carbs) that make it to your stomach in ingested beer are in low concentration compared to the carbs in the original mash, so are called residual. A typical glass of domestic beer will have about 14 grams of alcohol in it, and probably a little over 10 grams of carbs. Hence you get about 40 calories from the carbs, and 98 calories from the ethanol, giving a total of just under 140 calories, most of it derived from ethanol.

Drinking a single average beer will give you about the same number of calories as a can of soda or a thirty-five centiliter (twelve-ounce) bottle of sport drink, about 50 percent more calories than a glass of milk, and about five times the number of calories in a cup of coffee (with sugar and milk). Some beers have fewer calories than the typical domestic beer, and others have rather more; the least calorific hobbles in at 55 calories (Budweiser 55), while the most tips the scales at a monstrous 2,025 calories (Brewmeister Snake Venom). The typical bottle of beer, closer to our own preference, has enough calories (around 150) to fuel you for a forty-minute walk. So if you drink such a beer and go for a brisk stroll you might end up at a zero sum for calories, though if you stay home watching television you might only burn 15 calories. But of course, people's activity levels and metabolic rates vary wildly, so there is no one-size-fits-all formula here. For what it's worth, the daily zero-sum caloric intake for the average adult female is 2,000 calories, and for the average adult male, about 2,500.

The number of calories in a beer is, in general, proportional to the amount of alcohol in the beverage. High-alcohol beers tend to have high carbohydrate content. Brewers of light beers lower the caloric level by reducing the intended alcohol content, which in turn involves controlling the initial amount of sugar used in the brewing process. So if you had concluded that Budweiser 55 has a much lower alcohol content than Snake Venom you would be correct, although technically the caloric content of a beer is dependent on carbohydrate and protein concentrations as well as the amount of alcohol. Small sugar molecules should all have been converted into alcohol, so they will contribute little if anything to the caloric content of a beer. Depending on the hops variety, and how heavily the beer has been hopped, there will also be a slew of molecules and proteins from this source—but they will contribute a minuscule number of calories to the overall mix.

If we are to make sense of what happens to carbohydrates in our bodies, we must address the elephants in the room—obesity and being overweight. According to the Centers for Disease Control and Prevention (CDC), an individual is considered overweight if his or her body mass index (BMI) is between 25 and 29.5, and obese if the BMI is over 30. BMI has no units, and is rather a manipulated ratio computed using the equation:

$$BMI = \frac{\text{weight in pounds}}{(\text{height in inches}) \times (\text{height in inches})} \times 703$$

The CDC claims that BMI is a pretty good indicator of body fat content, and it is basically a standard for deciding whether an individual should be considered overweight or obese. Some nutritionists, however, suggest that BMI is not the best measure of overweight, and

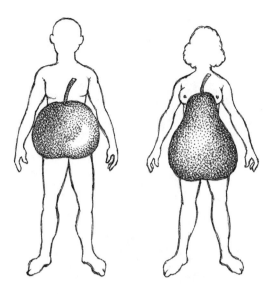

Figure 12.1. Apple- and pear-style fat deposition in men and women.

use instead the ratio of abdominal girth to hip girth. For men, this ratio should be 0.9, and for women it should be about 0.8. People with beer bellies typically measure out with ratios of 1.2 to 1.5.

Ultimately, all this excess fat comes from caloric content. Our bodies process a lot of the ethanol that is ingested, but there is also something else going on with calories from carbohydrates. The calories from the carbs you ingest with beer are used to provide energy needed for your everyday activity. If there are leftover carbs not immediately needed for energy, the body produces insulin to deal with the extra sugar in the bloodstream caused by those excess carbs. The small insulin molecule is a hormone that affects conversion of carbs into fat by regulating the level of proteins called lipases. These proteins break down fat molecules into fatty acids, which are then taken up by fat cells in specific regions of the body.

So where does the fat go in the body? The location of fat cells is different in men and women, which is why overweight men tend to be rounder when overweight, and women more pear-shaped (Figure 12.1). These cells will convert carbohydrates into fat if challenged. But they much prefer to absorb fat from food because carb conversion from

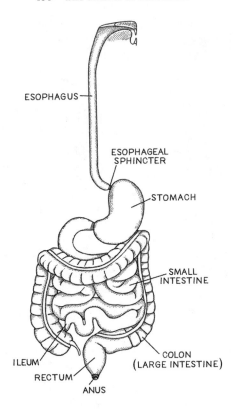

Figure 12.2. The human gastrointestinal tract.

fat costs ten times more in energy expenditure. Some researchers contend that this insulin system evolved as a product of natural selection during times of periodic starvation. This is the basis of the so-called thrifty phenotype. Thrifty individuals have the physiological capacity to store fat very efficiently and can easily become overweight in bountiful times—although they will get by better in times of shortage. Nonthrifty phenotypes use up their fat stores more rapidly and tend to be leaner, but will starve more readily.

When beer enters the mouth, its components start a circuitous route through the body (Figure 12.2). Once past the mouth, the beverage enters a part of the throat called the pharynx, then slides into the esophagus. Mucus lines both throat areas, and since beer is packed with proteins and enzymes, the digestive process begins there as enzymes in the mucus begin to break down some of the beer's components. Like many other molecules, ethanol is unaffected by the machinery of digestion, and passes by. It can, though, seep into the salivary glands of the mouth and throat, occasionally in concentrations high enough to damage those glands and thus impede the capacity to salivate. Ethanol is also toxic to some of the enzymes in the mucous layer of the esophagus.

The ingested brew travels the length of the esophagus, and finally encounters the esophageal sphincter, which leads to the stomach. If working properly, this sphincter will let in the brew while still enclosing the contents of the stomach. But ingesting large amounts of ethanol can cause the sphincter to be sluggish and allow backwash from the stomach to seep up into the esophagus. This seepage causes uncomfortable acid reflux, or heartburn.

Once in the stomach, the beer makes contact with some very strong digestive enzymes. Pepsin is a major player, and small digestive molecules such as hydrochloric acid are also present. A lot of ethanol can sneak through unscathed by these molecules, but other components of beer like carbohydrates and proteins are broken down. If ethanol is ingested in sufficiently high concentration it can disrupt normal stomach function, and may damage the organ by overstimulating the production of digestive enzymes. Any food present in the stomach will sop up some of the ethanol molecules, preventing them from doing damage and from entering the bloodstream.

The mash then passes on into the small intestine, where its components start to influence the sequestering of fat. Small molecules like ethanol and carbohydrates pass through the small intestine membranes, and flow into the bloodstream. The presence of carbs in the bloodstream triggers production of insulin in the pancreas and initiates the process that may lead to the storage of fat. If you don't burn the carbs off soon, the fat will accumulate in the fat cells of the body.

Many of us are personally familiar with the bulge above the waistline commonly known as the beer belly, and it is easy to fall into the trap of thinking that beer and other liquid libations with high caloric content are the cause of waist circumference increases. Indeed, Madlen Schütze and colleagues have shown that there is a 17 percent higher probability that your waist will increase in circumference if you drink beer. But it is more complicated than just a one-to-one correlation. Body weight and hip circumference are both involved—and Schütze and colleagues concluded that beer bellies are not entirely caused by the consumption of calorific liquids. Exercise and the capacity to burn off calories are also implicated. Like most aspects of human beings, every beer belly has a convoluted backstory.

A side effect of ethanol in the small and large intestines is to weaken the muscles of these organs, and to allow food to pass through relatively rapidly. The result is diarrhea, and disturbance of the microbiota of the gut. Scientists have known for a long time that our large intestines are filled with bacteria, and more recently have been able to determine the quantities of the different kinds of microbes that live in the long intestine, something that is apparently affected by drinking beer. In 2016 Gwen Falony and her colleagues examined the gut microbiomes of more than a thousand people using the DNA preserved in stool samples, much as DNA fingerprinting identifies the perpetrators in TV crime dramas. They showed that frequency of beer drinking had a big impact on the species of microbes living in the gut. Whether the difference makes beer drinkers more—or less—healthy overall is still unknown.

Molecules that reach the bloodstream are carried to the other organs of the digestive system, where they are further broken down for nutrients and energy. The two organs that are most active in this journey of the ethanol, carbohydrates, and proteins that come from beer are the liver and the kidneys. Kidney specialist Murray Epstein points out that our kidneys need a stable chemical environment, which ethanol disturbs. The kidneys control the body's water levels, along with those of several electrolytes such as sodium, potassium, calcium, and phosphate. If these electrolytes are disrupted, this organ system can get really upset. Too much ethanol in the kidney is, moreover, toxic to the antidiuretic hormone vasopressin, suppression of which tells the ducts

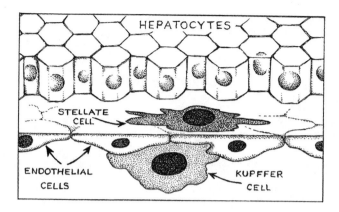

Figure 12.3. Diagram of liver cells. The Kupffer cells are immune cells that eliminate bacteria and other large objects. The hepatocytes do most of the work of the liver. Stellate cells and endothelial cells are also part of the liver's overall structure.

of the kidney to release water, diluting the urine that the kidney is producing. The diluted urine causes the electrolyte concentration in the bloodstream to rise, triggering the body to realize that it is dehydrated. This cascade of events is why it is a good idea to drink water when drinking beer, even though beer itself is mainly water.

The liver filters the blood, to rid it of toxins and other molecules that are not of use to the body (Figure 12.3). The filtering is done in subunits of the liver called lobules, of which the adult human body has around fifty thousand. Fine tubules throughout the liver act to increase the surface area of the lobules, increasing their exposure to the blood. The tubules are lined with cells of two kinds. Kupffer cells eliminate bacteria and other large toxic items, while hepatocytes, the workhorses of the liver, do a broad array of jobs that include synthesizing cholesterol, storing vitamins and carbs, and processing fats.

With respect to beer, the liver's most important function is to metabolize and filter out ethanol in the bloodstream. The more ethanol there is in the bloodstream, the harder the liver must work—and the greater will be the chances that some ethanol will not be filtered out and end up instead in the brain and other organs of the body. The liver's metabolism of ethanol is dependent on the enzyme called alcohol de-

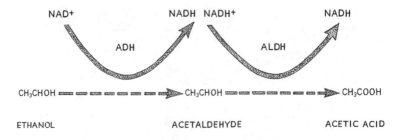

ETHANOL   ACETALDEHYDE   ACETIC ACID

Figure 12.4. Diagram of the breakdown of ethanol in the liver by the action of ADH and ALDH.

hydrogenase, or ADH (see Chapter 1). Ethanol is broken down by ADH into a molecule called acetaldehyde, plus a hydrogen ion. During the process a second hydrogen atom is sequestered onto a molecule inelegantly called nicotinamide adenine dinucleotide (NAD), to produce NADH. Acetaldehyde is toxic to the body, so must be degraded quickly. An enzyme called aldehyde dehydrogenase (ALDH) participates in this reaction, to produce acetic acid and an additional NADH molecule (Figure 12.4). Acetic acid is tolerated by our bodies, and is used by many organ systems as a source of carbons.

As discussed in Chapter 1, for most of our evolutionary history our ancestors were probably not ingesting much ethanol, suggesting that ADH and ALDH did not evolve in response to alcohol ingestion. Instead, these two enzymes were originally important in the metabolism of vitamin A (also known as retinol), and were later pirated away for the purposes of ethanol metabolism. Retinol and ethanol have similar shapes, so the enzymes work on both, facilitating this novel dual function. Using an enzyme called cytochrome P4502E1 (CYP2E1), liver cells also metabolize ethanol in a second way, oxidizing ethanol to acetaldehyde. This enzyme is normally not produced in large amounts. But when it is chronically bathed with ethanol the liver will kick into high gear, and produce a lot of it. Sadly, excessive amounts of CYP2E1 are associated with cirrhosis, a condition in which scar tissue starts to replace the normally functioning tissues of the liver. When cirrhosis occurs atrophy of the liver ensues, and the hepatocyte cells start to die. The liver becomes riddled with what are called Mallory bodies, and incurs massive and incurable damage—all of which makes an excellent argument for not overdoing it.

The two liver enzymes have been studied intensively with respect to their potential involvement in alcohol dependence, and it turns out that among human populations there is considerable variation in the genes controlling them. The ALDH variant called ALDH2.2 is found in high frequency in Asian populations (40 percent of people of Asian ancestry have it). The ALDH gene produces a protein that fails to efficiently break down acetaldehyde into acetic acid. As noted previously, aldehyde is toxic to our bodies; and if someone with the ALDH2.2 variant drinks a beer, acetaldehyde will accumulate in his or her tissues. An array of physiological responses results, the most visible being flushing of the face. People with this variant learn to shun alcohol because it causes discomfort and even pain. The CYP2E1 gene, too, has variants that have been implicated in alcohol avoidance. The protein this gene produces is active in the brain, and people with the variant phenotype become tipsy with very little alcohol in their systems. If they are smart, they stop drinking after the first couple of beers.

People with variant CYP2E1 and ALDH2.2 genes tend not to become alcoholics, for obvious reasons. Among the rest of the population, any tendency toward alcoholism seems to be determined in a very complex manner. In an attempt to decipher its genetic basis, scientists have used an approach called a genome-wide association study (GWAS), which allows comparison of whole genome sequences of hundreds of individuals both with and without alcoholism. The idea behind GWAS is that if the documented alcoholics in a study have similar changes in their genome, and differ in these respects from non-alcoholics, then those genomic changes can be correlated to the disease. Genome-wide association studies have been somewhat controversial, and results using this approach should be interpreted with care. What can be said, though, is that the tendency toward alcoholism appears not only to be controlled by many genes, but also to have a strong environmental component. This means that a precise genetic basis for the disorder may never be discernible. Certainly, the genetic causes of alcoholism continue to be mysterious. In practical terms, this means that everybody should be alert to the issue.

13

Beer and the Brain

We sat in front of two six-packs of the Spanish Er Boquerón beer, in a brave attempt to test the hypothesis that a hangover-free beer can be made. The bottle labels were emblazoned with the words "cerveza con agua de mar" because this beer is brewed using sea water, with the idea that the salt in the water will help prevent the dehydration that is a major cause of hangover. The beer itself is only 4.8 percent alcohol by volume, is slightly hoppy to the nose and palate, and drinks refreshingly. The six-packs disappeared quickly, and we are pleased to report that the next morning we felt no ill effects—though who knows what one more six-pack might have done? Another allegedly hangover-free beer can be obtained on draft at the De Prael brewery in Amsterdam. This brew includes several ingredients (including salt) that are not commonly used by brewers: ginger, vitamin B12, and willow bark. Each of these ingredients has a theoretical function in preventing the symptoms of a hangover, but their efficacy still needs to be tested scientifically to rule out a placebo effect. This research is on our agenda.

The Er Boquerón hangover-free beer contains about 5 percent alcohol, or about four teaspoons of pure alcohol per bottle. That is quite normal as beer goes. Once a beer has been swallowed, the digestive system breaks most of it down into particles that the body can use. During its passage, the alcohol in the beer will pass through several different organ systems; but some of it will nevertheless make it through untouched and find its way into the bloodstream. This will carry the surviving alcohol molecules to every part of the body served by blood vessels, including the brain. Since the brain is intricately laced with veins and arteries, alcohol hits many of its nooks and crannies. The proportion of those four teaspoons of alcohol that gets into the blood, and potentially to the brain, depends on many different factors. These include an individual's behaviors, as well as his or her genetic constitution. If you have just eaten a meal, less alcohol will get to the circulatory system because food particles in the stomach will have absorbed some of it. If you have inherited relatively weak versions of the enzymes that degrade alcohol, more alcohol will get there. Whatever the case, the effects will be fairly immediate. After your first beer, your blood alcohol level might be only around 0.02 percent: enough to give you a little buzz. And to understand where that buzz comes from, we need to know a little about the brain.

Human bodies and brains aren't designed to be exposed to very large amounts of alcohol, even though many humans tolerate it much better than the average mammal (see Chapter 1). Most of us can thus consider drinking six 4.8 percent ABV beers a pretty large amount. The main role of human physiology with respect to alcohol is to break it down and get rid of it, so that when we get drunk on beer—or on any alcoholic beverage—it is basically because the alcohol has defeated this breakdown system.

Once an alcohol molecule has reached the brain it can go pretty much anywhere the blood vessels reach, and because the body processes alcohol more slowly as we drink more beer, relatively more of it gets to the brain as our blood alcohol levels rise. The human brain is an amazing organ that has evolved from much simpler structures over hundreds of millions of years, and although evolution emphatically does not proceed by seeking perfect engineering solutions, a couple of engineering analogies are helpful in explaining how it functions.

Basically, there are two major problems that need to be solved if the brain is to take effective charge of the body. One of them is getting all the many different bodily structures to communicate with each other, and the other involves communication among the cells of which our tissues are made. Nature has solved the first problem by providing us with a nervous system that works a bit like the electrical circuits in our homes, with long, wirelike nerves linking the peripheral organs to the brain. The nerves coordinate between the brain and the other organs by propagating both chemical and electrical signals as adjacent nerve cells are stimulated by their neighbors.

An average human brain weighs about three pounds. If you were to hold one it would feel a bit like a handful of Jell-O, seeping slightly through your fingers. But you would also notice the many folds and wrinkles on its outer surface. Those folds are there because the external layers of the brain are rather like a thick cloth that must be scrunched up to fit inside the skull. The resulting creases result in the formation of gyri, which are the exposed parts of the folded sheet, and of sulci, which are the creases hidden inside by the folding. The cells in this crumpled sheet are interconnected, which makes the brain a great target for alcohol molecules. By the time you have downed the third beer of a six-pack, your blood alcohol will be up to around 0.05 percent, and you will be feeling pleasantly buzzed as the alcohol molecules begin to penetrate the farther reaches of your brain. You will also be more than halfway to the 0.08 percent at which most states will define you as legally drunk.

The simplest perspective on the human brain simply divides its cortex, the part covering the top and sides, down the middle between the right and left sides. Although many of the often-touted differences between the left and right sides of the brain have proven to be anecdotal, some of them really are significant. For instance, speech and speech comprehension are almost always centered in the left half of the brain. Still, this does not by itself make much of a difference to our understanding of how beer impacts the brain, because alcohol doesn't care about handedness, penetrating both sides of the brain with equal ease.

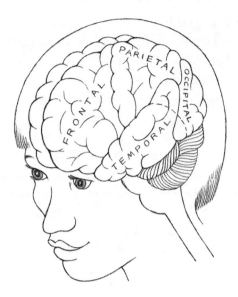

Figure 13.1. The four lobes of the brain: frontal, parietal, temporal, and occipital.

More important here is that each side of the brain is further divided into quadrants (Figure 13.1). This gives the exterior surface of the brain a total of eight lobes, each one susceptible to the effects of alcohol. At the front of the brain, under the forehead, is the frontal lobe. Just behind the frontal on each side is the parietal lobe, and below that is the temporal lobe. Finally, the occipital lobe sits at the back of the brain. The roles of these four major divisions of the brain are varied, even within each lobe. But each lobe also has a major function that we need to understand if we are to comprehend beer's effect on the brain.

The frontal lobes are where conscious decisions are made. The parietal lobes house the sensory and motor strips of the brain that are essential to our capacity for sensing and reacting to the outer world. Two important subareas of (usually) the left temporal lobe are named for the scientists who discovered them: Paul Broca and Carl Wernicke. They are involved in language speaking and comprehension, respectively. Finally, the occipital lobes at the back of the brain are responsible for visual processing and response. And the cells of every one of these lobes are exposed to alcohol molecules derived from the beer we drink.

Each of the four major lobes of the brain is made up of billions of cells known as neurons. These are connected to each other, forming pathways for information to travel along. Adjacent neurons communicate with each other via connections called synapses. What drives communication across synapses is the so-called action potential, a kind of electrical charge that courses from one neuron to the next (Figure 13.2). These electrical signals travel not only within the brain, but also outward, to carry instructions to various parts of the body and to transfer information back to the brain (see Chapter 11). Within the brain, they synthesize our perceptions of the outer world, and create our general sense of consciousness—though nobody knows precisely how they achieve this feat.

Another way to think about the structure of the brain is in terms of the color of the brain tissues: the famous white or gray cells (or matter). The white matter is made of billions of the neural cells called axons, which are insulated rather like electrical wires, and threaded throughout the inner layer of the brain. The outer area, or the gray matter, is also made up of billions of cells, but its organization is more complex than that of the white matter because of the presence of a second kind of neural cell, the dendrites. Dendrites connect to the nerve fibers (axons) of the white matter, by means of those synapses, and make a seemingly endless number of connections among themselves. The connections made in the gray matter are critical to the way the brain processes the data that the sense organs provide about the outer world. They are also essential for motor responses, as well as for memory, emotional response, and other higher-order neural functions. All this activity depends on the speed of the signals that the nerves carry, something that may be greatly affected by alcohol molecules.

A third way of looking at the brain is evolutionary. In this view, there are three parts of the brain, but this time going from inside to outside. Colloquially they are known, respectively, as the reptilian brain, the limbic system, and the neocortex, and they were acquired in succession over evolutionary time. Embedded deep inside, the reptilian brain includes the cerebellum, which is involved in both sensory and motor processes and controls our basic movements, and the brain stem, which controls our basic bodily functions. The limbic system is layered on top

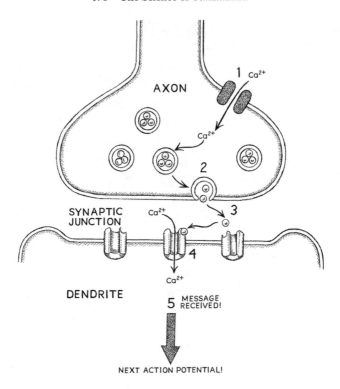

Figure 13.2. An overview of how a synapse works. A presynaptic axonal cell sits next to a postsynaptic cell (dendrite), separated by the synaptic junction. The membrane of the axonal cell has ion channels (1) distributed throughout it. Ions such as calcium are transported across the membrane, changing the ionic concentration in the presynaptic cell. This in turn releases small peptides (neuropeptides) from the cell (2). The neuropeptides then travel to the synaptic junction (3), where they interact with neural receptors (4) that are embedded in the dendrite membrane. As the neuropeptide binds to the receptor a channel opens, admitting more ions like calcium into the dendrite. This in turn triggers an electrical signal (5) that travels through the dendrite and on to the next neural cell.

of the reptilian brain, and is made up of many smaller neural clusters such as the hippocampus, the thalamus, and the amygdala, all important in both emotional and higher brain function. The reward center of the brain, which is hugely influenced by alcohol, also resides here. Finally there is the outermost cortex, in which higher reasoning takes place.

Brain cells need considerable nourishment (a 1.5 kg brain may use up to 25 percent of all the energy that a 100 kg individual consumes), and the neurons in all three parts of the brain are fed through an intricate network of blood vessels that very efficiently deliver oxygen—and alcohol—to them. By the time you are into your fourth beer, your blood alcohol content will be about 0.065, and you will be very close to being legally intoxicated. So let us move on to how those tiny alcohol molecules create what the famous Roman orator Seneca called *voluntaria insania* or "voluntary madness."

Envision the mass of neural cells communicating back and forth, from deep inside the brain to the surface, from one lobe to another, from the outer gray matter to the inner white matter, from the left side of the brain to the right, from distant parts of the brain to the limbic system, and among the various clumps of neural cells (nuclei) that are specialized to do specific tasks. There are around 100 billion of these cells, and each has the potential to make over 15,000 connections to other neurons. This means that there are about 100 trillion connections in the average brain. Even given differences caused by age (we lose synapses as we get older) and gender (females have fewer synapses than men), this is a phenomenal number of connections, far outstripping the number of stars in the Milky Way (a mere 400 billion).

Synapses in the brain and peripheral nervous system transfer signals from one cell to the next, allowing the information encoded in the signals to move to, from, and within the brain. If there were no control over how these signals are passed around, we would be a huge electrical mess. Proper control of the signals depends on the proper functioning of the synapses, and alcohol can have a huge effect here. Cells communicate with each other primarily via molecular interactions, and the action potentials across the synapses use ions as the currency of signaling. Those most commonly involved are sodium and calcium ions ($Na+$ and $Ca++$). It would be easy if action potentials could just jump from one cell to the next, across the two cell membranes. But alas, synapses did not evolve a simple and easy way of transferring electrical action potentials between the presynaptic cells from which they come and the postsynaptic cells that receive them. Instead, inside each presynaptic cell are vesicles that contain the small molecules known as neurotransmitters, while in the membrane of each postsynaptic cell are embedded hundreds or thousands of small proteins (Figure 13.2). Some of these protein molecules form little pores, or ion channels, through which chemical ions with electrical charges can pass. Others have highly specific structures that can bind neural transmitter molecules.

The ion channels are usually inactive until the presynaptic cell in which they reside reaches a critical concentration of ions. This concentration is caused by outside signals reaching the cell, for example from one of the sensory organs (tongue, eyes, nose, and so on). When the critical concentration is reached, the vesicles move toward the synapse and burst open, releasing the neural transmitters into the area between the two cells. The neurotransmitters then bind to the receptor molecules, and the bound receptors open the pores of the ion channels so that the ions can rush through. A new action potential in the postsynaptic cell is then created by the accumulating ions that have passed through the ion channels, and the neural transmitters become unbound from the ion channel proteins, shuttling back to the presynaptic cell in a process called reuptake, so that the whole process can begin again.

Our blood alcohol content is now up to 0.081 percent, and the alcohol from the five beers we have already drunk has been sneaking into these synaptic regions and begun to have some tricky effects. At first, the alcohol molecules caused a pleasant buzz. But as we continued drinking, the buzz gave way to weirdly euphoric feelings, and eventually to an increasing loss of physical control. What has been going on?

Well, the various neurotransmitters are critical controllers of action potential. There are over fifty different kinds of neurotransmitters, and each one has its own receptor. Depending on the precise neurotransmitter released into the synapse, a particular response will occur. On one hand, some neurotransmitters are excitatory, meaning they will make the synapses of the brain and nervous system more active. They will enhance the firing of action potentials and act as stimulants of the brain. On the other hand, some neurotransmitters are inhibitory, meaning they will impede action potentials so that the firing of synapses will slow and responses will be dulled. What's more, the rate of transmitter reuptake may also vary, and thereby change the rate of synapse firing.

Beer is a complex beverage, and its constituents may have subtle effects on the brain. On average, up to 95 percent of a brew is water. Alcohol is also present in various concentrations. Unfiltered beer will further contain yeast, and perhaps some bacteria. There will also be some by-products of brewing, such as phenols, alpha acids (humulones), beta acids (lupulones), pigment molecules, and any of many other products of fermentation. Besides the alcohol, a lot of these other compounds will eventually make it to the brain and have the potential to impact it.

Still, because the effects of alcohol are paramount, let's first see how this tiny molecule fares within the brain as a whole. One of the neurotransmitters that is impacted by alcohol is glutamate. This small molecule is an excitatory neurotransmitter, and normally enhances synapse activity and energy levels in the brain. When the alcohol content of the synapse is sufficiently high, it will cut down the amount of glutamate released by presynaptic cells and thereby slow the firing of synapses. This in turn slows down communication among the various brain subsystems, impeding coordination. On the inhibitory side, alcohol enhances the activity of the very important neurotransmitter known as gamma-aminobutyric acid, or GABA. This molecule inhibits action potentials, and hence slows down the synapses. Superficially this sounds a little like the effect that certain sedatives such as Xanax and Valium have on the brain, but it has been shown that alcohol acts in a different way from these sedatives. They increase GABA production, while alcohol increases the effect that GABA has on the synapse.

Alcohol is overall a depressant, so when you get drunk the tendency is to fall asleep. But alcohol also has a surprising stimulant effect that works via the reward center of the brain, which is located within the limbic system (Figure 13.3). Alcohol enhances the release of the neurotransmitter known as dopamine, whose levels are increased dur-

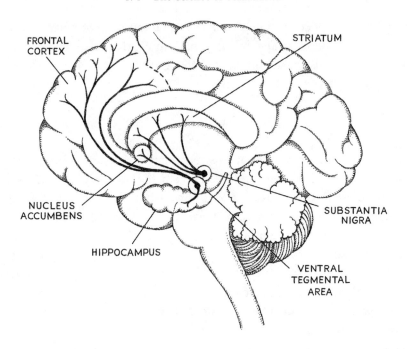

Figure 13.3. The reward center of the brain. The striatum is at the center of the brain's reward system, and interacts with several other brain regions as indicated in the diagram. These areas include the frontal cortex (where we make decisions), the nucleus accumbens, and the hippocampus (which modulates memory).

ing pleasurable activities, causing us to seek more of them. Alcohol thus tricks the reward center into craving more alcohol—even though the ultimate effect of that additional alcohol will be depression. It's a neurological catch-22: you drink more beer because your dopamine levels are going up, but as you drink that beer, your nervous system enhances those feelings of depression.

This scenario may make this an appropriate moment to down that last beer in the six-pack, which will get us up to a blood alcohol level of 0.093, well over the legal level for intoxication and a good point at which to take stock of the neural effects the beer has had on us.

After an entire six-pack of beer, most people feel a little sleepy and slow-witted. We slur our words, and may say inappropriate things. We have become very uninhibited, because the concentration of alcohol in our brain has begun to affect the prefrontal cortex in which we make decisions and modulate our behaviors. The alcohol has enhanced our GABA and dampened glutamate reception in our synapses, leading to slower firing of the prefrontal neurons. Of course, all that dopamine release might tempt us to begin another six-pack. But when we almost knock over a glass on the table, and realize we are losing coordination because alcohol has also invaded the cerebellum, we might decide this would be a bad idea. What's more, the balance organs in our inner ear are also getting bathed with the molecule, perhaps causing them to malfunction and make us think the room is spinning.

Because we are also feeling a little tired, we might decide to head home. This decision will have been mediated mainly by the general depressant nature of alcohol, and its impact on glutamate and GABA reception in all parts of the brain—most especially the brain stem, which has responded by slowing down many bodily functions, including breathing. It is the effect on the brain stem that signals we are sleepy.

If there is enough of that dopamine still lingering around, however, we might just decide to go ahead with that second six-pack. And in that case, we might well risk the hangover with which we began this chapter: an unfortunate condition that is caused not only by the dehydration that worries the brewers of Er Boquerón, but also by the dilation of the blood vessels of the brain, an effect caused by alcohol's tendency to lower the body's metabolism.

Even if you decline that second six-pack, there is no assurance that you will avoid the dreaded hangover. This is because alcohol also impacts the pituitary gland at the base of the brain. This tiny knob of neural tissue is responsible for making hormones that have all sorts of roles in regulating the proper function of your body. Too much alcohol will cause the pituitary to stop making the antidiuretic hormone vasopressin. This hormone tells your kidneys what to do, and by shutting down vasopressin production the pituitary signals your kidneys to send the water they produce straight to the bladder, bypassing the other parts of your body. So, even as your bladder fills up and you run more frequently to the bathroom, the lack of usable water elsewhere in your body is causing a whole slew of other undesirable effects. Water becomes extremely scarce in the body, and every organ selfishly rounds up any water it can. The brain isn't very good at doing this, so it suffers the most in the ruthless competition for water. It dehydrates, and consequently shrinks enough to tug on the connective tissues and membranes that separate it from the bone of the skull. It is this persistent tugging that causes the pain you experience during a hangover headache. And that explains the noble and never-ending search for a hangover-free beer.

PART FOUR

Frontiers, Old and New

14

Beer Phylogeny

Having finished putting together our beer phylogeny, we couldn't wait to try this Italian beauty, which amazingly combined all three of the major beer divisions we had identified. For inside the shapely brown bottle reposed a highly unusual blend of Belgian barleywine-style peated ale, Rauch Märzen aged in Scotch whisky barrels, and *Brett* beer. "Daydream," said the oversized cap. The head dissipated fast, as we'd expected from the *Brett* influence, and the dense, cloudy, yellowish liquid in our glasses gave off intensely aromatic odors of Cantal cheese, peat, and, of course, *Brettanomyces*. On the palate the *Brett*-y cacophony of flavors fascinated, while defying ready description. This clearly wasn't a beer for everyone, but it was certainly one to remember.

The human mind has a deep-seated need to make connections, and many readers will have seen t-shirts and posters decorated with wonderful diagrams representing the genealogies of the

different styles of beer. One of our favorites comes from popchartlab
.com. The poster in question is a true genealogy, because it shows ancestors and descendants, plus some cross-connections that make the
genealogy less treelike and more like a net. Although the sixty-five ales
and thirty-plus lagers in the chart represent the two big "families" of
beers, there is a line connecting the ales and lagers, indicating that there
existed a "grandmother" of all beer not indicated in the diagram. The
cousin-connections in the chart are particularly interesting, and include
Kölsch, cream ale, Altbier, California common, and Baltic porter, all of
which have characteristics of both ales and lagers.

Another chart, from bearingsguide.com, breaks out forty-five types
of ale and twenty-five of lagers. Like the popchartlab.com poster it has
cousin connections, but there are fewer of them and only cream ale and
Baltic porter are shown as difficult to place among lagers or ales. Some
beer genealogies, such as the cratestyle.com version, don't attempt to
show these cousin, or "hybrid," relationships. Yet other genealogies are
much simpler, and do not attempt to show relationships beyond the major
styles such as IPA or stout. Two of them (from Wikimedia Commons and
MicroBrews USA at https://microbrewsusa.wordpress.com/2013/07/17
/beer-family-tree/) don't even try to connect lagers with ales in their diagrams, and give only the bare outlines of beer relationships. And one of
our evolutionary biologist colleagues, Dan Graur, prefers a more treelike
diagram, with no connections between ales and lagers.

Why are we mentioning these essentially decorative diagrams?
Well, in attempting to visually display the relationships among beers,
the creators of these posters have tweaked our professional nerve endings. As scientists who have spent a collective seventy years worrying
about the relationships among organisms like archaic humans, lemurs,
fruit flies, bacteria, plants, and what have you, we see some of the
beauty of phylogenetic analysis in these posters, as well as some of the
trickier challenges that face our field.

Systematists, the scientists who try to sort out the kinds of organisms in the biosphere and the relationships among them, have always
used phylogenetic (evolutionary) trees to represent those relationships.
Probably the first such tree was published by the French natural historian Jean-Baptiste Pierre Antoine de Monet, Chevalier de Lamarck, in

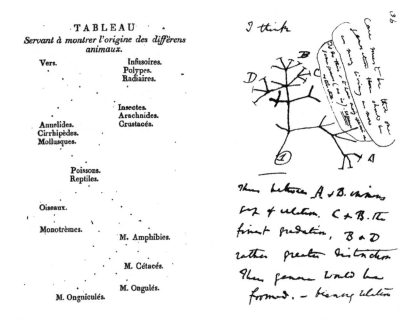

Figure 14.1. *Left:* Lamarck's 1809 "tree" figure. Note that "M. Amphibies" resides at a node of the tree, and is seen as a transitional form between "M. Cétacés" and "Poissons, Reptiles." Darwin's "I think" tree (*right*) clearly implies that the taxa on the tips of the tree are living, and that the nodes in the tree are ancestors.

1809 (Figure 14.1). Because Lamarck was also the first scientist to suggest that life changes over time, it is perhaps not surprising that he did something we avoid doing today: he placed living taxa at the nodes (ancestral points) of the tree, suggesting that some living groups had transformed into others. This is something that all the beer posters do too—which is okay for a genealogy, but not for an evolutionary tree, in which ancestors are hypothetical, or at best known as fossils.

In 1836, in a private notebook, Charles Darwin followed up with his famous branching "I think" tree (Figure 14.1). This was an attempt to express diagrammatically how evolution proceeds, using hypothetical taxa (organisms) throughout. It is clear, though, that he considered the taxa at the branch tips to be living ones, and the nodes themselves to be ancestral. Darwin went on to formalize the notion of the phylogenetic tree as an explicit statement of ancestor-descendant relationships in his 1859 book *On the Origin of Species,* and indeed it was he who coined the poetic metaphor the Great Tree of Life.

Phylogenetic trees have been used for a long time in the study of the evolution of organisms. They are enormously useful in evolutionary biology for many reasons, including that they show us how specific entities are related to others, and how ancestors fit into the picture of evolution. So, confronted with a diversity of beers that is—to beer lovers at least—as glorious as the diversity of life itself, it occurred to us that it might be instructive to use systematic techniques to look at how beers evolved. Of course, beers didn't evolve in the same way organisms did, but it turns out that quite similar patterns may be produced by evolution in both the cultural and biological domains. Indeed, linguists have long used trees to represent the relationships among languages, often constructing them using techniques that are remarkably close to those invented by biologists.

The beer genealogies we see on those posters and t-shirts are based on immense knowledge of the products concerned. In this respect they closely resemble the way in which evolutionary trees were typically constructed half a century ago: by specialists who used their extensive expertise to arrive at them intuitively. This sort of thing could not go on indefinitely, and during the 1960s a new generation of systematists began to complain that this procedure was unscientific, based on feelings rather than on actual data. They began to seek more objective alternatives.

The 1960s was a tumultuous decade in general, and systematic science was no exception, with lots of infighting and sometimes vituperative disagreement. Eventually, three basic approaches to tree-creation emerged, all of them still used to some extent today. One approach simply asks how similar organisms are to one another in various respects, and uses the sum of those similarities to make the tree. You take all the pairwise differences (the flip side of similarity) between the species you are interested in, and ask which difference is the smallest.

The pair that is least different is then placed on the first node of the tree. The next-most-similar species to the first two is then placed as the closest relative of that grouping, and so on. This way of proceeding is called the distance method, and it differs from the other two in compacting all the information that one might have for the species into a single similarity (or distance) measure.

The other methods take each of the different bits of information (called characters, or character states) about the organisms in the analysis, and assess how well the individual characters tell an evolutionary story. Both methods are concerned with tree shape (topology), and aim to assess how well the characters fit onto all possible arrangements of the species in the analysis. In the maximum parsimony method, the tree that best fits all the characters used is accepted as the simplest and thus the best explanation for the data. The maximum likelihood method also proceeds character by character and looks at all possible trees, but it uses probability to decide which tree is the best. This method requires that you have a prior model for how the characters evolved (that is, you assess the probability of having the data at hand, given both the tree and the model). It is relatively straightforward to develop prior models for how molecules change, but it is considerably more difficult for anatomical structures. So, because the characteristics used to categorize beers are more comparable to anatomy, we will turn our backs here on maximum likelihood, and concentrate on the maximum parsimony approach.

Let's look at maximum parsimony a bit more closely. Say we want to "systematize" three kinds of beers: an American lager, a Belgian IPA, and a Vienna pils. Any beer aficionado will know the result of this exercise already, but please bear with us because knowing the method of analysis is important. And the first thing we need to understand is that a tree with just three beers in it is meaningless until we decide where the root—the ancestry—of the tree lies.

A tree with those three beers in it would look like the one on the left in Figure 14.2. It is hard to argue that there is much information in the tree as it stands. But if you root the tree on any one of the three branches, as seen on the right of the figure, it immediately becomes apparent that the other two branches are closely related to each other. The

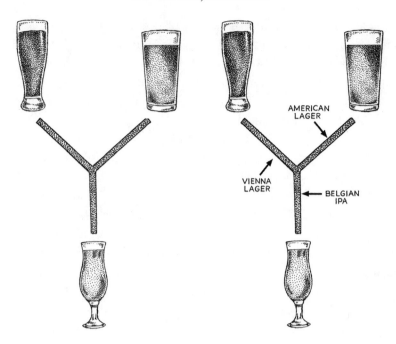

Figure 14.2. The unrooted tree for three beers: Vienna lager (VL), American lager (AL), and Belgian IPA (BI). The three potential rooting positions for the three-beer problem are shown with arrows. If the tree is rooted at AL, then Belgian IPA and Vienna lager are each other's closest relatives. But if the root of the tree is at BI, then Vienna lager and American lager are each other's closest relative. The third possibility is that the root lies at VL, which would mean that American lager and Belgian IPA are each other's closest relative.

position of the root, in other words, is critical to the evolutionary information it conveys. But there are many ways in which to root any tree. One could do as we just did, and arbitrarily choose one branch, but this is hardly objective or even repeatable: someone else might equally come along and say, "I think the root is over here instead," and repeatability is out. Choosing purely from your own expertise will not, in other words, give you a good root. The only way forward, then, is to add a fourth beverage, and root the tree on that one. This approach is called out-group rooting, and it requires a beer that is outside (less related to) the three "ingroup" beers being examined. In this exercise, a wheat wine would fill the bill nicely.

The next step is to generate a character matrix for our four beers. This character matrix is the heart of the analysis, and it contains all the information we have that might be useful. If we were doing this kind of analysis for a group of organisms we would look at them from top to bottom, trying to characterize their behavior as well as sequencing their DNA, as demonstrated in Chapters 7, 8, and 9 for barley, yeast, and hops, respectively. With beer DNA isn't going to help us, so we will need other information for our three ingroup beers and our outgroup. This is the fun part, and with just three beers to sort out we could easily sit down with a bottle of each and get gustatory characters from drinking them. But since there are more than a hundred major kinds of beer out there, we will need to take a shortcut.

Fortunately, the Beer Judge Certification Program (BJCP) produces a document that outlines the specific characteristics of a hundred or so styles of beer. The BJCP document includes most of the information about these beer styles that we need to construct a matrix for phylogenetic analysis. Of greatest importance are what the BJCP calls tags. These describe characteristics of strength, fermentation approach, color, region of origin, style, family, and dominant flavor. For example, color has three states: pale, amber, or dark. The BJCP also includes original gravity and final gravity, international bitterness units, alcohol by volume units, and a more quantitative color measurement called SRM (standard reference method). By combing through the BJCP guidelines, one can collect information on some twenty or so characters that can be useful in constructing a phylogeny of those hundred-plus beers.

More characters can be obtained from recipes for specific beer types. Fortunately a database also exists for this, in the form of Beer Smith.com, a website that archives thousands of recipes for different styles of beer. This database also gives us characters like the taste rating, the kind of yeast and barley used, and more specific fermentation information.

For our example, we will look at three of the tag characters to see how the parsimony approach works, and to show how we ultimately go about constructing our beer phylogeny. Let's use these six BJCP tag characters: strength (very high, high, standard, or session), color (pale,

Table 14.1. CHARACTERS AND CHARACTER STATES
USED IN THE THREE-BEER TREE WITH OUTGROUP

	strength	color	yeast	locale	style	taste
Wheat wine	high	amber	top	North America	craft	balanced
Belgian IPA	high	pale	top	North America	craft	hoppy
American lager	standard	pale	bottom	North America	traditional	balanced
Vienna lager	standard	amber	bottom	Central Europe	traditional	balanced

amber, or dark), fermentation yeast (top or bottom), locale (North America, Central Europe, Eastern Europe, Western Europe, British, or Pacific), style (traditional, craft, or historical), and dominant taste (balanced, hoppy, sour, or bitter). For the three ingroup beers and the single outgroup beer we get the character states in Table 14.1.

Next, to make the analysis a little easier, we recode the character states. The search for the best tree with only three ingroup beers and these six characters is fairly straightforward; but as more beers are added the number of possible trees grows exponentially, so that the one hundred or so beer types we will want to analyze eventually means that we will need to look at over 10^{100} different trees (a one with one hundred zeros after it). This is a task that only a very large computer can take on. And while we could program the computer to handle character states like "pale" and "bottom," it is much easier to give those states numerical values so that the computer can handle them more easily. Accordingly, for strength, standard is recoded as "0" and high is scored as "1"; for color, amber is recoded as "0" and pale is recoded as "1"; and so on. Our matrix then looks like the character states in Table 14.2.

Next comes the really serious part of the phylogenetic analysis—testing how well the characters fit onto the possible trees. We already saw that the broad analysis of beers we want eventually to do would require looking at 10^{100} different trees; fortunately, the number of trees we need to examine for three ingroup beers is just three, all shown in Figure 14.3. The wheat wine serves as the outgroup.

Table 14.2. RESCORED CHARACTERS AND CHARACTER STATES
USED IN THE THREE-BEER TREE WITH OUTGROUP

	strength	color	yeast	locale	style	taste
Wheat wine	1	0	1	1	1	1
Belgian IPA	1	1	1	1	1	0
American lager	0	1	0	1	0	1
Vienna lager	0	0	0	0	0	1

Let's look at how the six characters sit on the three trees shown in Figure 14.3. For the strength character (Figure 14.4), our outgroup is scored as "high" and to map strength onto the AL + VL tree on the left will take a single change, from high to standard, just before the node connecting AL and VL. For the AL + BI tree in the middle, though, the number of changes required is two: one change on the AL branch above the AL + BI node, and one on the VL branch. Similarly, the VL + BI tree on the right requires two steps, one on the VL branch above the VL + BI node, and one on the AL branch. If strength were the only character we were looking at, we would conclude that the AL + VL tree is the most parsimonious because it takes a single step while the other two involve two steps. The yeast and style characters yield similar numbers, so we actually have three characters that support the AL + VL tree. Color, however has a different pattern: the VL + BI tree has a single change, while the other two trees have two steps each. Locale and taste are both what biologists call phylogenetically uninformative, because they can be mapped onto all three trees with the same number of changes, and hence don't help us decide which tree is better.

There are two ways to look at the three-beer problem at this point. First, there are three characters that agree completely with the VL + AL tree, there is a single character that agrees completely with the VL + BI tree, and there is no support for the AL + BI tree. The second way of looking at the problem is that the VL + AL tree requires five changes, the VL + BI tree takes seven changes, and the AL + BI tree takes eight

Figure 14.3. The three possible trees for the three-beer problem. The tree on the left implies that American lager is closest to Vienna lager. The tree in the middle implies that American lager is closest to Belgian IPA, and the tree on the right implies that Belgian IPA is closest to Vienna lager. The best solution to the problem is to add an outgroup. Since the outgroup is the same for each of the three trees, it is not shown.

Figure 14.4. Mapping the "strength" character onto the three possible beer phylogenies. The white bars indicate where on the tree a change from "high" to "standard" is required. The phylogeny on the left requires a single change, occurring on the branch leading to the AL + VL node. The other two trees both require two changes, as indicated. Note that this is the same pattern as for the "yeast" and "style" characters.

changes (if we ignore the phylogenetically uninformative characters). Either way, the VL + AL tree wins the competition and we can declare it the most parsimonious.

This relationship, of course, makes sense with respect to what we already know about these beers. Both the lagers are—naturally enough—lagered, both use bottom-fermenting yeast, and both are also traditional brews. The most parsimonious tree also tells us that color changes twice, suggesting that color might not be such a good trait to use to answer this particular phylogenetic question. The reason for this is what biologists call convergence. The phenomenon of convergence is extremely interesting in evolutionary biology, providing the proverbial exceptions that prove the rule: similar features (convergences) may evolve independently in different lineages, simply as a response to similar problems. Birds, bats, pterosaurs, and some insects all have wings, but this is not because they are closely related and inherited their wings from a common ancestor. It is because they all fly.

When we move to analyzing the full matrix with 103 styles of beers (the major families and styles included in the BJCP, and essentially the styles found on the popchartlab.com genealogy), the analysis naturally becomes a bit more difficult. First, what do we use as an outgroup? This is a tough question, because if we choose an outgroup that is too far away (milk, say), the root of the tree will become random and meaningless, and if we choose an outgroup from within the beers (barleywine, for instance), we will run the risk of artificially rooting the tree next to it. As a compromise, we tried rooting our tree with two relatively closely related beverages, gruit and wine. Next, as already mentioned, came the sheer computational difficulty that is involved in evaluating 10^{100} different trees. Computing the solutions for so many trees poses what mathematicians and computer scientists call an NP-complete problem: even though we know there is a finite solution to the problem, we don't have the computing power to find it, so we must seek another road. In other words, with this many trees to examine we need to use shortcut methods that eliminate large swaths of those trees that will definitely not be part of the solution.

One shortcut method of rooting the tree would be simply to interpose a root between the two larger groups of beers (say, lagers and ales) into which all the other beers are thought to fall. But then we would risk falling into the expertise trap, assuming what we are hoping to discover. Before rooting the tree in this way, we would thus want to confirm the

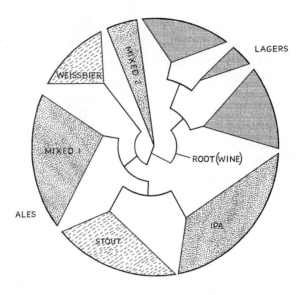

Figure 14.5. Phylogenetic tree of beer, obtained by rooting with wine. The three discernible "monophyletic" groups of beers in the ale clade are stouts, IPAs, and Weissbiers. Our splitting of the lagers into three groups is discussed in the text.

true unity of each group. And a final caveat before we dive into the analysis: with the limited number of characters we have available for our beer phylogeny, changing our way of analyzing the tree (removing characters or beers from the analysis) will produce different trees. This doesn't mean that the approach is unsound, but it emphasizes how important it is to remember that the assumptions that go into the analysis are critical to its outcome.

In the end, we stuck with the BJCP tag characters, and did two analyses: one with gruit as an outgroup, and one with wine as the outgroup. The scoring of wine is difficult because it is produced so differently from beers; ultimately more than half of the characters had to be scored as ambiguous or missing. This sometimes happens in phylogenetic analysis, especially when incomplete fossils are included, but fortunately computational means have been developed of dealing with this problem. In the end, both analyses showed lagers and ales as forming good and distinct groups. We present the tree rooted with wine in Figure 14.5. This shows the lagered beers arising from one single node in the tree, whereas the ales arise from another. This confirms the idea of a deep split between lagers and ales—and it could also, by the way, justify rooting the tree between ales and lagers, without any outgroup. The emanation of ales and lagers from independent single nodes recalls the biological phenomenon of monophyly, whereby all species within a particular group are descended from a unique common ancestor.

Finding ales and lagers to be monophyletic confirms earlier ideas about the homogeneity of these groups. There are also some satisfying similarities between our two differently rooted trees for the ales. For example, IPAs, stouts, sour beers, historical beers, Belgian ales, American ales, and Weissbiers all form pretty good monophyletic groups. But while ales form a large monophyletic group in both trees, there are some differences between the two trees for ales other than those IPAs, stouts, and Weissbiers. Since there are many ways to categorize styles of beer, differentiation beyond major categories is difficult to determine from this analysis. But amid this uncertainty we can still count twelve major groups of ales, some of them diverse, while others embrace at most three styles.

Our phylogenies depart in certain respects from the genealogies shown on those t-shirts and posters. For example, the popchartlab.com genealogy shows three major divisions of lagers: American, German, and pilsner. Our phylogeny implies four groups—international/American lagers, Czech lagers, bock/dunkel lagers, and pilsners. One oddity involving lagers is the inclusion of Kölsch in this group. This is surprising because Kölsch is a top-fermenter and is not lagered. It appears, then, that Kölsch has converged on lagers in respect to the various beer traits we have used, something that would explain its apparently transitional position in the popchartlab.com diagram. The two other purportedly transitional forms—Baltic porter and cream ale—also have interesting placements in our beer phylogeny. Baltic porter, a bottom-fermented darker beer, sits firmly in the phylogenetic tree within the bock/dunkel group, casting doubt on any transitional status. The top-fermented and pale-colored cream ale, by contrast, is the first beer out from the lagers, and its location in the tree suggests that it does indeed lie in a transitional position.

There are other ways to depict beer relationships than construct-

Figure 14.6. STRUCTURE analysis of 104 beer styles, using 5 populations (K=5). The five groups that emerge are IPAs, stouts, lagers, and two heterogeneous groups of ales, the first including Belgians, goses, and lambics, and the second including Scottish, Irish, and bitter beers.

ing genealogies or trees. There is, for example, a Periodic Table of Beer Styles that clusters beers, and suggests relatedness among them through the proximities of different styles in the table. We also saw earlier that some evolutionary researchers prefer to use non-tree-based approaches to look at relatedness among organisms. The STRUCTURE approach is used widely in evolutionary studies, and if clustering is what we want to do, then principal components analysis will be useful too (see Chapter 5). We tried applying both these latter approaches to beer, using the same database as for our phylogenies.

Our STRUCTURE analysis of the 103 beer styles (and gruit, to make 104) suggests that there are five populations (K=5), as shown in Figure 14.6. The five groups are IPAs, stouts, lagers, and two heterogeneous groups of ales: the first including Belgians, goses, and lambics, and the second embracing Scottish, Irish, and bitter beers. Of additional interest are the brews that cannot be typed to a single group. On the far left in the diagram, these include American amber and American brown beers. The outliers of the Belgian-gose-lambic group are also heterogeneous, and include American pale ale, blonde ale, Weissbier, and weizenbock. California common and Belgian dubbel appear as odd beers within the Scottish-Irish-bitter group. Among the lagers, the hangers-on are cream ale and pre-Prohibition lager. Stouts appear to be very easily typed, while in contrast American brews have some deviant characteristics that make them difficult to place unambiguously into a specific group. In other words, they have some major aspects of, say, IPAs; but they have also picked up some characteristics from other

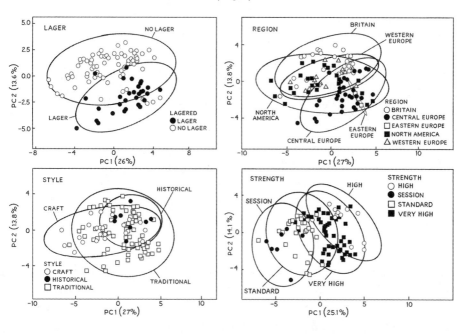

Figure 14.7. Principal component analysis of beer, centered on lagering (*upper left*), style (*lower left*), region (*upper right*), and strength (*lower right*).

groups. It is interesting that cream ales are placed in the lager group (albeit with some ambiguity), because they are not lagered.

The principal components analyses (Figure 14.7) are more difficult to interpret, because the clusters overlap in many aspects. To show how little distinctive clustering there is, we generated PCA analyses centering on lagering, style, geographic location, and strength. Lagers and ales cluster separately as expected, with little overlap, confirming the results from the other kinds of analysis. Many of these observations are in good agreement with other ways of grouping beers, but what is very interesting is that there is no overall agreement with *all* of them. Where beers are concerned, things are apparently always complicated.

We should note that this is not the only time that PCA has been used in the context of beer. Marketers and advertisers have used it to examine the preferences of different categories of consumers, and we will probably see it used more widely as brewers and distributors try to refine their knowledge of the markets they are serving.

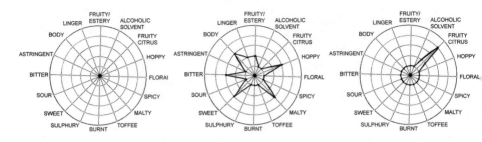

Figure 14.8. Taste wheel (after 33books.com) showing the sixteen categories of beer taste we scored for over fifty beers from southern Germany. An unmarked wheel is shown on the far left. The middle shows the wheel for a beer we enjoyed, and the right-hand diagram shows one we could barely finish. All beers tasted will remain anonymous.

Phylogenetic analysis can be done on any sort of data. But it is only meaningful under certain assumptions. To explore this further, we traveled to the Czech Republic and southern Germany for a beer-tasting field trip (which just happened to coincide with the opening of Oktoberfest in Munich). Our goal was to taste as many beers as possible in a seven-day period, and to classify those beers using phylogenetic methods. Instead of using the published style guidelines we had used in our previous phylogenetic tree, we used a different method of characterizing the beers we tasted, one that employs a variation of Morten Meilgaard's taste wheel discussed in Chapter 11 (Figure 14.8). The particular taste wheel we used came from 33books.com, and we highly recommend it if you want to keep a record of the beers you've tasted. In this wheel, the characteristics of a given beer are scored clockwise from noon, for the following categories: fruity/estery, alcoholic/solvent, fruity/citrus, hoppy, floral, spicy, malty, toffee, burnt, sulfury, sweet, sour, bitter, astringent, body, and lingering. The various characteristics were scored from one to five, and both of us had to agree on the score before it was entered. As an example of our scoring we show in Figure 14.8 a beer that we really liked, and one that we could barely finish (though bravely we did).

The taste wheel is a neat visual way of expressing a taster's reaction to a beer, and it quickly became obvious to us that the spikier the final wheel looked, the more we liked the beer concerned. We also

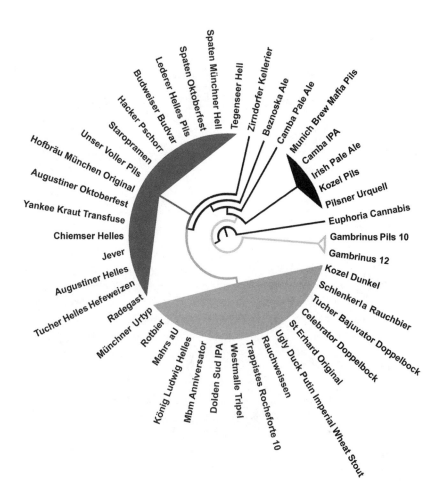

Figure 14.9. Tree of beers we tasted in southern Germany in 2017. After scoring the fifty or so beers with the taste wheels, we analyzed the data using parsimony. We rooted the tree with a *Cannabis* beer. Two major groups of beers emerged using this approach. The dark gray clade, or group, holds many of the Oktoberfest beers we tasted. The light-gray clade included beers that we preferred, with strong tastes. One smaller clade holds beers that did not impress us (solid black clade), and a two-beer clade uniting the major tasty pils also exists (white clade).

Figure 14.10. Our t-shirt design, in general agreement with the phylogenetic results obtained in this chapter.

noted that we could easily take the beer-wheel scores and transform them into a phylogenetic matrix for constructing a tree of the beers we tasted. Here is the scoring for two of the beers tasted, with their characteristics listed as if clockwise from noon:

Kozel Dunkel   1 1 1 2 1 1 4 2 3 1 4 1 2 1 2 3
Redgast   1 1 4 2 1 2 1 1 1 1 2 1 2 2 2 3

We then entered those scores into a phylogenetic analysis that gave us the results shown in Figure 14.9.

We were gratified to discover that the branching diagram we obtained quite closely reflected our subjective preference not only for the individual beers we tasted, but also for the styles involved. We were surprised at how neatly the analysis separated the beers into ones that we had enjoyed (light gray group) and the ones we merely tolerated (dark gray group). Interestingly, most of the pils, helles, and Oktoberfest beers that we tasted were in the "merely tolerated" category. All of them were good beers, but the beers we preferred were all darker, hoppier, and in general smokier. One odd group was made up of an IPA, three pilsners, and a pale ale, all of which appeared to us to be quite bitter. Two beers at the base of the tree, both pilsners from the Gambrinus brewery in Plzeň, were quite amazing. We also felt that three other beers—two ales and a Kellerbier—were likewise exceptional. Their position apart from all the other beers in the tree is most likely because we recognized them as having unusual but admirable tastes.

We started this chapter with a description of posters and t-shirts that depict beer genealogies. Now we end it with a design for our own t-shirt (Figure 14.10). Rather than splitting the various styles of beer into subgroups, we prefer a taxonomy of beer that embraces five major groups of ales (stouts, Belgians, Weissbiers, Scottish, and IPAs) and three major groups of lagers (darks, ambers, and pales). The categories of beer recognized by the BJCP are then allocated to these groups, based on the phylogenetic tree in Figure 14.5. We also recognize two "hybrid" events: one involving Kölsch, the other cream ale. But since there is no disputing matters of taste you will, of course, have your own ideas.

15

The Resurrection Men

We opened the bottle and poured the yellow topaz liquid into our glasses, where it sparkled slightly. On the nose it yielded unfamiliar fragrances, and we might not have guessed it was beer. On the palate, though, this delicious beverage was all flowers, peaches, and honey, and the finish lingered forever. The recipe for this brew had been imaginatively re-created from the analysis of chemical residues coating the insides of ancient metal drinking vessels excavated from a Phrygian tomb of the late eighth century BCE that might have been the last resting place of the legendary King Midas. If so, he lived well, that Midas.

The origins of the American craft beer movement lay in the rejection of industrialized brewing. And since a closely related desire was to explore brewing's ancient artisanal roots, if not necessarily to return to them permanently, it was only a matter of time until someone tried to replicate an ancient beer. This goal was not as straightforward as it might seem: the only physical evidence we have of early beers comes in the form of chemical residues left behind in pottery jars after the liquid contents vanished, and those residues offer only a faint reflection of the original ingredients and chemical complexities of the long-evaporated beer (see Chapter 2). But there is some documentary evidence for ancient beer-making as well, so it is not surprising that the first American attempt to resurrect a beer from the ancient world involved Ninkasi, the fourth millennium BCE Sumerian goddess of beer celebrated in the ode we cited earlier.

In 1989 Fritz Maytag, a wealthy young entrepreneur who had not long before bought and rejuvenated San Francisco's venerable Anchor Brewing Company, ran across a 1987 article by the Philadelphia anthropologist Sol Katz. In this piece Katz argued that gathering cereal grains for brewing had been a major impetus for the agricultural revolution; and in support of this contention he cited the translation of the *Hymn to Ninkasi* that the University of Chicago Assyriologist Miguel Civil had made a couple of decades earlier. Working closely with Katz and Civil, Maytag then figured out a practical recipe for Ninkasi's beer that was compatible with her activities as described in the *Hymn*. He brewed and bottled a batch, which came out at a respectable 3.5 percent ABV, and presented it at the annual meeting of the American Association of Microbrewers.

Those fortunate enough to be in attendance sipped the archaic brew from large jugs, through long drinking straws "fashioned to resemble the gold and lapis-lazuli straws found in the tomb of Lady Pu-abi at Ur." Seven months later, the remaining Ninkasi bottles were opened for a group that convened in Philadelphia at the University of Pennsylvania Museum of Archaeology and Anthropology. Although it had been kept under refrigeration, a lot of the Ninkasi beer had gone off— but what survived was pronounced "similar to hard apple cider," with a "dry taste lacking in bitterness." The museum's Patrick McGovern described it as having "the smoothness and effervescence of champagne and a slight aroma of dates."

It is hard to imagine that, however unconsciously, modern brewing expertise had not somehow intruded into the making of an ancient beer as good as these encomiums suggest. But then again, despite the railings of spoilsports who claim that it cannot be conclusively proved that the Babylonian product contained any alcohol at all, maybe the worshipers of Ninkasi really were drinking a pretty respectable brew. Certainly, it is difficult otherwise to understand the ancient poet's abounding enthusiasm for the stuff: "While I circle around the abundance of beer / While I feel wonderful, I feel wonderful / Drinking beer in a blissful mood."

Various other brewers, both in the United States and in Europe, subsequently took a stab at making ancient beers. In Egypt, for example, dry conditions have made for exceptional preservation of brewery sites, including the fragile raw materials. During the 1990s an archaeological team led by Cambridge University's Barry Kemp excavated several ancient breweries at the Middle Kingdom site of Amarna, where the archaeobotanist Delwen Samuel identified traces of malted barley (or maybe emmer wheat) that had been heated, then sieved to remove the hulls. There was no evidence of honey or date juice, leading Samuel to suggest that the hieroglyphic symbol usually translated as "dates" in ancient Egyptian beer recipes may in fact have simply meant "sweetness," of the kind derived from the malt itself. All this suggested a brew very different from the Sumerian one, and Kemp went to the people at Edinburgh's Scottish and Newcastle Breweries to ask if it could be re-created.

After careful consideration, a Scottish and Newcastle team under Jim Merrington brewed a 6 percent ABV beer from malted emmer wheat. It was flavored with coriander and juniper, but had no sweeteners. A thousand bottles of this Tutankhamun Ale were made, selling for a fabulous price at the famous Harrods department store in London. Reportedly the beer was a hazy gold in color, and tasted "fruity, grainy, with caramel/toffee, sweet/spicy astringent, and with a dry finish." Scottish and Newcastle made no more—and alas, since 2009 the company itself is no more. But others would get in on the act. In 2010, to celebrate the local opening of the traveling Tutankhamun archaeological exhibit, Denver's Wynkoop Brewery produced a Tut's Royal Gold using ingredients widely available in Egypt. The fermentable base was furnished by a pale barley malt, wheat, the short-stemmed local cereal teff, and honey; tamarind, coriander, grains of paradise, orange peel, and rose petals were used as flavorings. There are no archaeological indications to suggest that this exact recipe was ever used in ancient

Egypt, but somehow the combination seems to fit the spirit of the effort. And just in case it's not extreme enough for you, you could also try the same brewery's Rocky Mountain Oyster brew, which comes in at 7.2 percent ABV, and a lavish 3 BPB (you figure it out).

In Scotland, the Williams Brothers Brewery not only occupies the historical William Younger's brewery buildings in Alloa, but also produces a range of ales inspired by ancient archetypes. Most famous of these is its Fraoch Heather Ale, a gruit ale claimed to be in the tradition of the brews first documented at Skara Brae some 2,500 years ago. Fraoch is made by introducing sweetgale and flowering heather into the boiling malt, which is then left to cool for an hour in a vat with fresh heather flowers before fermentation is started. In a modern flourish, Fraoch is matured in used oak casks previously employed to age sherry and malt whisky. The result is a deep amber ale with whisky as well as herbal aromas, and a sweet barley wine finish. Satisfying a draught as it may be, drink one of these and you are tasting a tradition, rather than an accurate replica of an early gruit.

Many others have by now also tried their hand at making ancient beers, and the Internet overflows with advice for any home brewers who may be tempted to do so. But no one has approached the re-creation of ancient beers with quite the tenacity, vigor, expertise, and concern for sheer authenticity (tempered by knowledge of the tastes of modern beer drinkers) shown by the biomolecular archaeologist Patrick McGovern and his collaborator Sam Calagione, founder and head brewer at Delaware's Dogfish Head Brewery. McGovern is the world's leading expert on the composition of ancient fermented beverages, and Calagione is by common consent one of the most creative and interesting craft brewers in the United States. In the late 1990s the duo embarked on an ambitious attempt to bring back to life ancient beers whose traces have been found at archaeological sites all over the Old and New Worlds: a venture entertainingly recounted, complete with recipes for beer (and, for the epicure, food), in McGovern's fascinating book *Ancient Brews*.

The enterprise began when McGovern was asked to analyze the chemical residues in metal pots found by members of a University of Pennsylvania Museum of Archaeology and Anthropology expedition at the ancient site of Gordion, in central Turkey. During classical times Gordion had been capital of the Phrygian kingdom that, in the late eighth century BCE, was ruled by a king Midas who was probably the legendary monarch with the golden touch. A burial mound at the site proved to contain an unopened central burial chamber that held both the remains of a male who had died at the age of about sixty-five, and a large collection of drinking paraphernalia whose style suggested an eighth-century BCE burial date. Apparently, the various cauldrons, jugs, and drinking bowls used at the funerary feast for the tomb's royal occupant (either Midas or his father, Gordius), and the leftover food and drink had then been buried beside him, in their original containers, to give him a good start on the journey to the afterlife.

A quarter of the bronze vessels found in the tomb proved to contain yellow residues from evaporated ancient beverages. Using an array of scientific equipment, McGovern and his colleagues demonstrated that these residues contained tartaric acid. In Turkey this compound is most commonly found in grapes, suggesting that some sort of wine had originally been present. Beeswax compounds also betrayed the former presence of honey, and the tests finally identified beerstone, evidence of barley. The results were consistent from vessel to vessel, suggesting that all had held a mixed fermented beverage with elements of wine, mead, and beer. An extreme beverage, indeed.

That knowledge was fine for archaeological purposes. But when it came to reconstructing this ancient drink, a lot of questions remained unanswered, as McGovern relates in *Ancient Brews*. What were the proportions of the principal ingredients? What had contributed the intensely yellow color to the residue? Were the ingredients prepared separately and then mixed together, or were they all brewed together? Where did the yeast come from? What kind of cereal had been used, and what kind of honey? What kind of grapes, and fresh or dried? And was the final product carbonated? Without answers to all these questions, and others, no one could ever be certain that any re-creation would entirely or even closely resemble its original. Still, the ultimate success of any brewing enterprise has always—then as now—been due to the intuition and craft of the brewer, which is every bit as important as any of the inert ingredients.

The initial re-creation of the king's funerary beer resulted from a challenge McGovern presented to craft brewers who assembled at the University of Pennsylvania Museum in the spring of 2000 for the yearly tasting event with the beer and scotch critic Michael Jackson. Calagione finally won the challenge with a beverage that incorporated incredibly expensive but intensely yellow saffron as the bittering agent. He also used Greek thyme honey, Muscat grapes, a mead yeast, and a two-row variety of barley. Still, using the three known basic ingredients—ill-matched as they might seem to modern people who tend to store meads, beers, and wines in separate compartments of the mind—Calagione managed to come up with a pale golden-yellow beverage, with a heady 9 percent alcohol by volume, that was aromatic and well balanced. It had a sweetish attack, with biscuit and honey flavors, and it finished clean and dry. It became an immediate cause célèbre. A descendant version is still on sale, nearly two decades later, as Dogfish Head's Midas Touch.

Emboldened by this success, McGovern and Calagione next moved to re-create the world's oldest known beer, chemical traces of which had been found at the nine-thousand-year-old site of Jiahu, in north-central China. There, pottery jars that had contained liquids had also proved to retain residues that both resembled and differed from their equivalents in the Midas jars. Just as at Gordion, beeswax testified to the former presence of honey, and tartaric acid to either hawthorn fruit or grapes, or both (in China three times as much tartaric acid is present in hawthorn fruit as in native grapes). But while the third main component was a cereal, it was not barley: it was rice. Because of its complex composition, McGovern chose to call the resulting beverage not a beer, but a Neolithic grog (he had also called the Midas beverage a Phrygian grog). He suggested that the presence of those multiple sugar sources indicated the ancient makers wanted to raise the alcohol content of their product as much as possible, as well as to provide other sensory delights; he also pointed to the documented use in ancient and modern China of a wide range of herbs and microorganisms to encourage saccharification and fermentation.

Grog or not, McGovern turned again to the beer brewer Sam Calagione and his team to reconstitute this ancient drink. After various experiments that included an attempt to convert the rice starches to sugar using mold cakes (*qu*) like those used in making rice wine, the group eventually settled on a protocol that featured four basic ingredients: hawthorn fruit (dried and powdered, for legal reasons), Muscat grapes, orange blossom honey, and a gelatinized rice malt (with hulls and bran). All four ingredients were then brewed together, initially using sake yeast but switching to an American ale yeast after several fermentations became stuck. No herbs were added, since it was only a speculation that the Jiahu brewers had used them. Twelve days of fermentation raised the alcohol level to 10–12 percent; the resulting mixture was then conditioned in the tank for four days at room temperature, and for forty-six days at a reduced temperature.

There was a lot of guesswork involved, but McGovern is confident that the resulting Chateau Jiahu bottling was a reasonable facsimile of what the ancient Jiahu people had tasted. When you open a bottle (not an option for the Jiahu folk) and pour the beverage into a glass, you will first notice the deep yellow color, and the formation of a champagne-like light mousse on the surface. This is followed by a sweet-and-sour flavor profile that McGovern correctly says is an ideal accompaniment for Chinese cuisine. Chateau Jiahu went on to win prizes, and McGovern claims it is the favorite among all his many re-creations. But again, modern expertise had substituted for all the unknowns in the original brewing process, and we will never be sure quite how closely it resembles the stuff that evaporated out of those clay pots nine thousand years ago.

McGovern and Calagione went on to replicate other ancient beers from around the world. By McGovern's own admission they pushed the envelope with Dogfish Ta Henket (bread beer), which tried to capture the spirit of the long-lived ancient Egyptian brewing tradition. The brewers used residues from three different sites and periods to come up with a recipe that included barley malt, doum palm fruit, emmer wheat bread, za'atar (a spice mixture), and chamomile: the entire mix was then fermented using yeast recovered from fruit flies captured in a date palm oasis. As a generic representation of a long and diverse brewing tradition it was probably fairly close to the mark, but alas the pungent fruit and strong herbal flavors of Ta Henket proved not to have the wide public appeal of the fragrant Chateau Jiahu.

Even more extreme was a beverage that never made it to market: a chicha maize beer, based on the beverage that fueled the Inca Empire. Today, Quechua-speaking Peruvians make numerous versions of this brew by sprouting corn in a process analogous to the malting of barley. But in ancient times the same trick was accomplished by extracting the sugar in the corn using salivary enzymes—in other words, by "chewing and spitting" (see Chapter 2). In 2009 McGovern and colleagues made a heroic attempt to produce a chicha in this time-hallowed way, chewing their way through a pile of red Peruvian corn for eight hours, then mixing the masticated remnants with a puree of pepper berries and wild strawberries. "So we would not be accused of poisoning the populace," McGovern wrote, "we made sure to boil the mixture." Fermentation was done using a standard American ale yeast, and the deep crimson 5.5 percent ABV beer was ready in time to celebrate the launch of McGovern's book *Uncorking the Past* on October 8, 2009. Attendees at the Great American Beer Festival that year lined up to taste what was left. Later versions were, alas, made with soursop fruit as the sugar source.

Over the years, McGovern and Calagione have re-created several other ancient (and extreme) beers, including the popular Theobroma, based on a Mayan chocolate beer, and a Viking kvasir that features both wheat and barley malts, cranberries, lingonberries, honey, birch sap, bog myrtle, and yarrow. Because so much in the beer-making process is not preserved in the raw list of ingredients that chemical analyses can (at least partly) detect, and written records are poor to non-existent, these beers have all had to be made in the spirit of the original, rather than as exact replicas. This exercise has resulted in some highly interesting brews, without which the world would certainly be a poorer place; but to get a complete sense of the original we would need an explicit written recipe, a luxury unavailable for any truly ancient beer. Even the elaborate *Hymn to Ninkasi* is woefully lacking in critical details.

We are also short on information about some historically important beer styles that have lapsed or changed so that, even if the name survives in some form, nobody knows any longer exactly how the original was made, or what it tasted like. This certainly applies to one of today's best-selling craft styles: India pale ale. As we saw in Chapter 3, IPAs came to prominence as strong, malty, and hoppy English ales that were miraculously improved by the rigorous journey in sailing ships around the Cape to India. Although colonial and other beer-drinkers were ecstatic about it, the IPA style eventually waned as the nineteenth century progressed: both changes in fashion and crude taxation policies led to its replacement by weaker and less characterful brews. By the turn of the twentieth century, only older beer drinkers could even remember consuming it. Enter the English beer writer Pete Brown, who around the turn of the twenty-first century wanted to know what the classic IPA had been like, and how its flavor profile had been transformed by its arduous voyage. To answer these questions, he had a beer made in the original IPA style and took it via ship to India, an undertaking engagingly recounted in his book *Hops and Glory*.

Brown could embark on this adventure only through the generous cooperation of colleagues at what remained of the famous Bass Brewery of Burton-on-Trent in central England. IPAs were invented in London, but the focus of IPA brewing later switched to Burton, and Bass eventually became the largest producer. The Bass archives still contained the detailed recipe for Bass Continental, an IPA in a late style that had been exported to Belgium in the 1850s, and this formed the basis for a replica IPA that was brewed using the oldest equipment available. Aromatic Northdown hops went in, along with pale and crystal malts, two traditional Burton strains of yeast, and, perhaps most essential, gypsum-rich Burton well water. After tank-conditioning for several weeks, five gallons of the resulting ale were drawn off into a small cask to be taken to India. According to Brown, when fresh from the tank the brew was deep amber, with a "massive nose of tropical fruit-salad aromas." But then there was a "bitter, resinous spike" on the palate, and the finish just "tailed off." The reconstituted Bass Continental was hardly an ideally balanced ale at this stage, but no matter; what was important was how the brew would taste when it arrived in India.

The tale of how it got there was one of crisis and misfortune. Most significantly, the original barrel exploded in a hot apartment in the Canary Islands, and had to be substituted with a small metal keg. Hastily filled by his co-conspirators from the same batch of ale, this container joined Brown's ship in Brazil. The bulk of the journey across the South Atlantic and around the tip of Africa remained to be accomplished at this stage, and the keg and its guardian reportedly experienced a lot of bucketing-around on its long sea journey to Bombay and then overland to Calcutta, the seat of the East India Company at the time IPA was first brewed. There, four months and many thousands of miles after it had entered the fermentation vat, the beer had hardly had a chance to settle before the keg was tapped, squirting pressurized foam far and wide. Not the best of auguries, perhaps, but as Brown tells the tale:

> It poured a deep copper colour, slightly hazy from the sheer weight of the hops. The nose was an absolute delight: an initial sharp citrus tang, followed by a deeper tropical salad of mango and papaya. [Then] my tongue exploded with a rich, ripe fruit, seasoned with a hint of pepper. That bitter, hoppy spike had receded, the malt reasserting itself now against that hoppy attack . . . There was a delicate tracery of caramel . . . the finish was smooth and dry, clean and tingling. And by God it was damned drinkable for its hefty 7 per cent alcohol.

As Brown himself admitted, after all his many trials and tribulations he might well have been predisposed to render a favorable judgment on his IPA, but his companions at the tapping ceremony were apparently equally enthusiastic. Moreover, comparing his accounts of the ale at the beginning and end of its epic voyage leads to two rock-solid conclusions: first, the journey really had transformed the beer; and second, the result of this reenactment was delicious. Clearly, in beer as in everything else, progress and improvement are not always synonymous.

16

The Future of Brewing

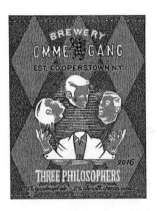

In an era when creative microbrewing is flourishing in the lengthening shadow of the megabrewers, predicting where beer and beer-drinking are going is tricky at best. So we decided to leave it to the combined brainpower of the three philosophers whose images decorated the eponymous bottle that reposed in front of us. Leavened with a touch of Belgian kriek, the upstate New York quadrupel poured a deep chestnut brown, with a creamy head and a mere hint of those cherries in the highlights. The malty cascade of flavors was perfectly balanced, and the finish could only be described as unctuous. A true beauty. The cheerful lesson of this hybrid beer seemed to be that, even as brewing moves into an uncertain future, the ingenuity of the brewers themselves will continue to be irrepressible.

Before you try to predict the future, it is always a good idea to consider the past. And in the case of beer, the recent past has been very eventful. For on both sides of the Atlantic, two separate but opposite trends have intersected, each originally formed in opposition to the other.

When Prohibition ended in the United States, brewing briefly blossomed. But boom quickly turned to bust, and many smaller brewers did not survive or were snapped up by the emerging giants (see Chapter 3). Beer became a commercial commodity, and by the 1970s only a handful of big brewers remained. Then the infamous beer wars erupted as Anheuser-Busch of Saint Louis went on the rampage, with an advertising budget and a ferocious determination that almost no rival could match. Among the national brewers, only Milwaukee's Miller contrived to withstand the onslaught, initially because it was owned by the wealthy tobacco concern Philip Morris, and later because it cleverly capitalized on the "light beer" craze that began in 1975 with the introduction of Miller Lite, a beverage inspired by a particularly insipid German lager. Advertising genius turned an unassertive product into the most sought-after beer on the market, and a lawsuit-punctuated standoff ensued between Miller and its aggressive Saint Louis rival, even as such magic names as Schlitz, Pabst, and Rheingold were falling by the wayside. By the 1980s only one other major brewer, Coors, had somehow managed to survive. Soon, the three brewers controlled 80 percent of the American beer market.

Something similar was happening on the other side of the Atlantic. In England, filtered, pasteurized, and carbon-dioxide-pressurized keg ales had been introduced just before World War II. Initially intended for export, these beers proliferated on the home market during the postwar years because of the advantages they offered to those who made, distributed, and sold them. The more interesting cask-conditioned ales from which the keg beers were descended demanded—and still demand—a lot of work, not only on the part of their brewers, but also from the publicans, who have continually to tend them in the cellar and flush out the pipes through which the brews are drawn up to the bar by beer engines. A new postwar generation of landlords, and the breweries for whom they nearly all now worked, had found it much simpler to serve, and transport, the blander and mostly rather weak new keg ales. What is more, those ales could now be nationally branded and distributed, a development that encouraged the merging of regional breweries into national colossi, eventually known in the United Kingdom as the Big Six. By the mid-1970s, keg ales were accounting for more than half of pub beer sales, even as the number of different brands of beer available was cut in half between the mid-1960s and the mid-1970s. Almost as bad for lovers of tradition, easy-to-distribute-and-serve bottled beers also began to gain ground in pubs as well as in supermarkets.

Adding to the problem, British brewers had always thought of themselves as producers of beer. But along the way they had developed significant real estate interests by way of the vast numbers of pubs and prime brewery sites they now owned. Predators were not going to ignore this juicy target forever, and at the beginning of the 1960s Canada's Carling, a major lager producer, began an acquisitions spree that sparked the eventual emergence of the Big Six, presaging the end of big brewing as an avocational enterprise that sat above the general commercial fray. Brewing concerns increasingly became the target of, and were folded into, metastasizing multinational behemoths that viewed beer exactly as they would any mass-market commodity.

In Britain, this globalization of brewery ownership first allowed market penetration by both bottled and draft mass-market lagers, and then, through massive advertising, spurred the popularity of these beers among a new generation of drinkers. Although domestically produced, the new lagers were promoted and sold mainly under international brands, in a development that truly turned the British beer market upside down. Long a bastion of ales, Britain rapidly became a nation of lager drinkers. By 2014 polls showed lager to be the choice of some 54 percent of British beer quaffers, although this number has slipped somewhat since. Meanwhile, the trend toward consolidation in the beer industry has continued unabated. In 2008, even America's gigantic Anheuser-Busch was itself swallowed by the Belgian-Brazilian brewing colossus InBev. As if this were not enough, in October 2016 the combination absorbed SABMiller, by then the world's second-biggest brewer as the result of multiple mergers, including with Coors. To satisfy regulators, a diminished MillerCoors was spun off from the new behemoth, but this did little to encourage diversity within the American brewing industry.

The trend toward more globalized beers everywhere was bound to provoke an eventual reaction. In Britain, disappointed ale drinkers began to form pressure groups advocating for the revival of cask-conditioned beers. Foremost among these was the Campaign for Real Ale (CAMRA), which was established under a slightly different name in 1971. Founded by a group of journalists who had spent their formative years protesting against nuclear weapons and practically everything else, the new organization laid into the giant brewers with gusto, organizing boycotts, staging mock funerals for small breweries that had gone under, holding regional and national beer festivals, and publishing annually the influential *Good Beer Guide*. Inventing the term "real ale" was, as the adman and beer writer Pete Brown has pointed out, a stroke of marketing genius. The big brewers were taken aback and were forced to respond as sales of keg ales plummeted, and the regional brewers were reinvigorated. Thanks to CAMRA agitation, the priceless heritage of British real ale continues to flourish to this day; by its reckoning, some 1,500 U.K. breweries currently produce real ale.

Nobody this side of Wall Street would dispute that the revival of real ale was a wonderful development. But the one ironclad rule of human experience is the Law of Unintended Consequences. In harking back to tradition and the past, the real ale movement in Britain had a braking effect not entirely unlike the one the *Reinheitsgebot* had in Germany. There can be no question that, in deploring any relaxation of brewing standards, and emphasizing the purity of tradition, the *Reinheitsgebot* had played a major role in maintaining the quality of German beer over the centuries. But at the same time, rule-bound tradition tended to stifle innovation in a craft that begs to evolve. Although most widely distributed German beers were, and are, beautifully made, there is a certain uniformity among them: in mainstream German brewing, there has traditionally been only one kind of perfection.

That said, there were always alternatives: wheat beers have been perennially popular in Germany, and quirky local traditions such as those smoke beers from Bamberg continued to flourish alongside the pure products of the *Reinheitsgebot*. Possibly because of this safety valve, the German beer-drinking public has remained remarkably satisfied. No German CAMRA-equivalent has emerged, and there has been none of the popular discontent that dislodged poor Ludwig from his palace in Munich two hundred years ago. Nonetheless, there was clearly an unrealized demand for something more creative, because after European Union rules forced the adoption of a liberalized brewing law in 1993, a dynamic and hugely interesting German craft beer scene developed.

The efforts of CAMRA to revive real ale in the United Kingdom had another unintended consequence: many beer drinkers began to look back to what they had lost, but not in sufficient numbers to oust the lagers relentlessly promoted by the international beer conglomerates. As a minority interest, cask-conditioned ales soon became identified with a subculture of eccentric, backward-looking enthusiasts who were sufficient in number to keep tradition alive, but not numerous enough to entirely reinvigorate the market. A sort of standoff between lagers and ales thus developed in British pubs. But we may be approaching a tipping point. For, quite aside from the British beer-nerd tendency that thankfully remains alive and well, there are now hopeful signs in the wider market that real ales are clawing back lost ground from the bulk lagers.

Across the Atlantic, the situation in the 1970s was completely different. After all, in the United States there was no tradition to revive, Prohibition and its sequelae having basically killed off not only local beer brewing but also the social beer-drinking customs that had once been centered on ale houses and taverns. Ice-cold industrial beer, drunk at home straight from the refrigerator, had taken over. But in a country brimming with creativity and entrepreneurialism this could not last forever; the United States was primed for its own craft beer revolution.

Most historians trace the origin of this revolution to Fritz Maytag's takeover of San Francisco's failing Anchor Brewing Company in 1965, and to his efforts in the years following to resurrect traditional beer-making styles. Yes, that's the same Fritz Maytag who re-created Ninkasi's beer—and who also brewed the first American IPA in 1975.

President Jimmy Carter signed legislation legalizing home-brewing in 1978. Before long, many of the new home brewers had turned pro, and the craft beer revolution was truly under way—although exactly what a craft beer is remains unclear. The sternest definitions of the term refer to small scale of production, and require adherence to traditional practices of beer brewing, with no adjuncts (such as non-barley-malt sources of sugar) or artificial ingredients—an admonition frequently honored in the breach. Some definitions additionally stress independence (from the big brewers), although, as we will see, this distinction is beginning to erode. Stylistically, almost anything goes in this realm: craft brewers make porters, stouts, pale ales, sour beers, and even—or especially—those extreme beers we've described. The most extreme among them use almost anything imaginable that might be fermented, and—should you wish to include them in the category—they make nonsense of any definition of craft beers that bans adjuncts. What's more, some craft beer makers do not even make their own product, instead contracting the actual brewing of their beers to larger concerns that can afford the equipment they do not yet possess. Craft brewing is thus an avocation or industry that is still finding its identity, and a craft beer is something you basically know when you see it.

An important early influence in craft brewing was Jack McAuliffe. In 1976 his New Albion Brewery in Sonoma was the first brand-new American brewing operation to open in a very long time, in a country that was much more accustomed to brewery closures. Realizing that he could not compete head-on with the lager giants, McAuliffe decided to create a niche market in flavorful ales and porters, and to stress beer as a civilized accompaniment to food. The New Albion enterprise was enormously influential among the local cognoscenti, but alas it was not a financial success. Like many pioneering endeavors, it eventually went under without ever having achieved the scale required to be profitable: its capacity was in the order of four hundred barrels a year, whereas in

1976 Anheuser-Busch had several breweries across the country, each producing more than four million barrels annually.

Still, it was in the spirit of New Albion that more business-oriented brewing entrepreneurs like Jim Koch, who founded the Boston Beer Company in 1984, finally made inroads into the sales of the beer behemoths. Ironically, Koch built his market share by contracting out the actual brewing of his "Sam Adams" beers, and by concentrating his energies and finances on marketing. But once he had established his own brewing operation, and Sam Adams had become arguably the country's leading "craft" brand (Boston Beer produced 2.3 million barrels in 2013), Koch teamed up with McAuliffe to brew a re-creation of the latter's legendary 1976 New Albion Ale.

Meanwhile, other pioneers like Oregon's Fred Eckhardt, author of *A Treatise on Lager Beers,* and Ken Grossman of California's Sierra Nevada Brewing Company, were spearheading the rapid proliferation of craft beers across the country. Charlie Papazian's influential establishment of the Great American Beer Festival, first held in Boulder, Colorado, in 1982, helped this growing trend. And while the pioneering beer writer Michael Jackson's influential 1988 *New World Guide to Beer* did not address the U.S. craft beer movement specifically, in praising Maytag's Anchor Steam Beer it helped pique curiosity worldwide about the sheer variety of beers becoming available in the United States. At last, American consumers were being made aware that there was much more to beer than the products of the megabrewers, and that there were some interesting alternatives out there.

This awareness showed up in the marketplace. By 1985 there were already thirty-seven craft brewers in commercial operation, and in the following decade the number rose exponentially. Then the industry faltered: from 1,625 U.S. craft breweries in 1998, the number fell to 1,426 in 2000, mainly due to problems of quality control as production expanded dramatically. But after a decade of recovery the number of American craft brewers began to rise again, from 1,750 in 2010 to 2,418 in mid-2013. As of 2018 there are over 5,000, with more than 20,000 individual labels and 150 self-identified styles.

By the second decade of this century, the craft brewers, despite their tiny beginnings, were also beginning to make serious inroads into the sales of the brewing leviathans, whose market shares were stagnating. Public preference was undergoing a discernible shift from lighter lagers to darker, more flavorful styles. From a 2 percent share of the market in 1995, craft beer reached 6.4 percent in 2012; today, the craft share of the U.S. beer market is reckoned to be around 10 percent and rising.

This trend is hardly something the behemoths can ignore, and they have responded in two ways. One has been to launch their own "craft" brands. For example, MillerCoors (as it is now) markets its "Belgian White" beer under the Blue Moon label, without taking any pains to reveal its own involvement. Blue Moon is actually a pretty good product and has done quite well (selling over a million barrels annually). By contrast, few beer drinkers have heard of Anheuser-Busch's Elk Mountain or SAB Miller's Plank Road labels.

The second strategy has been to buy successful craft businesses. Anheuser-Busch bought into Seattle's Redhook Brewery as early as 1994, and purchased Oregon's Widmer Brothers Brewery three years later. Although they remained separately operated, both acquisitions were promptly ejected from the Brewers' Association that speaks for the craft beer industry. Widmer already had a stake in Chicago's respected Goose Island Brewery, and Anheuser-Busch InBev bought out the rest in 2011. Unsurprisingly, Goose Island lost its official craft beer status when this happened, as did three other leading craft breweries that have recently sold out to AB InBev.

The trend continues. In 2015 half of the Californian icon Lagunitas went to Heineken, the world's third-largest brewer (which has subsequently snapped up the rest), and San Diego's Ballast Point Brewing sold out to the Constellation Brands wine-and-spirits conglomerate. Several craft brewing operations across the country have recently gone to private equity investors, and some craft breweries have even joined forces with private equity firms to buy out floundering smaller rivals. In this way, Big Brewing on the one hand, and the kinds of outside economic forces that resulted in the formation of Britain's Big Six on the other, are gaining a stronger foothold in an industry niche whose vitality depends on the agility, creativity, and commitment of those innovative entrepreneurs who have made this such a memorable period for beer drinkers in the United States.

With more than five thousand breweries in the United States alone (surpassing the 1873 pre-Prohibition record of 4,131), and with virtually every beer-drinking country enjoying its own energetically diversifying craft beer scene, commercial brewing is certainly ripe for a shake-out. But what will the future industry look like? Most of those craft breweries make at best a few thousand barrels annually, and in the current highly competitive market, most will not be viable for the long haul without significant mergers and consolidations. It remains to be seen just how those consolidations will be achieved. If the megabrewers use their financial muscle and unparalleled distribution channels to swoop in and sweep up the cream of the crop, there is some reason to fear a return of uniformity, despite the majors' professed commitment to quality. After all, their core competence is the massive-scale production and distribution of a perfectly consistent product, so the temptation for the giants will always be to see quality and uniformity as more or less synonymous. The dependability of product the big brewers achieve is admittedly a minor miracle of chemical engineering, but historically, achieving it has not always operated in the interests of those beer drinkers who see virtue in diversity. Nostalgics lament the fate of Bass pale ale after the company fell into the hands of Global Beer, and purists whisper that even the legendary Pilsner Urquell, now owned by Japan's massive Asahi (via SABMiller and AB InBev), is not quite what it once was. Nonetheless, the big brewers certainly know a marketing opportunity when they see one and have clearly understood that it is in their interests to maintain a certain diversity of product.

The wholesale engulfing of craft brewing is, of course, a worst-case scenario. Although the big brewers will always have an important presence, the craft brewers have already shown that they have a substantial niche. If the inevitable pruning within that niche takes place mainly through mergers among craft brewers—mergers that redistribute talent while creating more economically viable brewing facilities and distribution chains on a subindustrial scale—the outlook for those who love beer and prize its variety will be a cheerful one. At the very least, one can reasonably hope that craft beers will continue to flourish alongside their mass-marketed counterparts. By some estimates, craft beers will soon have won over 20 percent of the beer market in the United States and worldwide; though it seems likely that a significant part of that fraction will wind up being controlled in some way by the brewing giants, it seems improbable that the megabrewers will ever take over entirely. One survey found that 44 percent of American Millennials have never tasted Budweiser—though, alarmingly, their tastes in alcoholic beverages may be straying toward spirits on the one hand, and nonalcoholic beverages on the other.

So while beer drinkers almost everywhere currently have more choice in terms of style and concept than they have ever had before, it is clear that they are enjoying this bounty within a brewing environment that is in transition. Fortunately, it is they who will have the last word. For it is above all the consumer who will determine whether the coming trends in brewing will hark back to the bland uniformity of the past, or continue along the present path of taste-expanding innovation and variety. Aware and discriminating drinkers are beer's best guarantee of a diverse and interesting future.

Annotated Bibliography

There is a vast popular literature on beer, much of it centered on home-brewing. Here we provide a chapter-by-chapter annotated list of the major technical and popular published sources consulted in the writing of this book, including all sources of quotations.

Chapter 1. Beer, Nature, and People

Tyson (1995) invented the phrase "The Milky Way Bar" for the extraterrestrial cloud of ethanol molecules. Wiens et al. (2008) reported the drinking habits of pen-tailed tree shrews, and Schoon, Fehr, and Schoon (1992) the fate of the drunken hedgehog. For a general discussion of ethanol consumption in nature, see Levey (2004). See Dudley (2000, 2004) for the drunken monkey hypothesis as well as alcohol aversion in mammals, and Milton (2004) for an alternative view. For alcohol and fruit flies, see Starmer, Heed, and Rockwood-Sluss (1977), Shohat-Ophir et al. (2012), and Milan, Kacsoh, and Schlenke (2012). Alcohol dehydrogenases in primates were analyzed by Carrigan et al. (2014); chimpanzee tippling was reported by Hockings et al. (2015).

Carrigan, M. A., O. Uryasev, C. B. Frye, B. L. Eckman, et al. 2014. "Hominids Adapted to Metabolize Ethanol Long before Human-Directed Fermentation." *Proceedings of the National Academy of Sciences of the United States of America* 112: 458–463.

Dudley, R. 2000. "Evolutionary Origins of Human Alcoholism in Primate Frugivory." *Quarterly Review of Biology* 75: 3–15.

———. 2004. "Ethanol, Fruit Ripening, and the Historical Origins of Human Alcoholism in Primate Frugivory." *Integrative and Comparative Biology* 44: 315–323.

Hockings, K. J., N. Bryson-Morrison, S. Carvalho, M. Fujisawa, et al. 2015. "Tools to Tipple: Ethanol Ingestion by Wild Chimpanzees Using Leaf-Sponges." *Royal Society Open Science* 2: 50150. http://dx.doi.org/10.1098/rsos.150150 (accessed June 7, 2018).

Levey, D. J. 2004. "The Evolutionary Ecology of Ethanol Production and Alcoholism." *Integrative and Comparative Biology* 44: 284–289.

Milan, N. F., B. R. Kacsoh, and T. A. Schlenke. 2012. "Alcohol Consumption as a Self-Medication against Blood-Borne Parasites in the Fruit Fly." *Current Biology* 22: 488–493.

Milton, K. 2004. "Ferment in the Family Tree: Does a Frugivorous Dietary Heritage Influence Contemporary Patterns of Human Ethanol Use?" *Integrative and Comparative Biology* 44: 304–314.

Schoon, H. A., M. Fehr, and A. Schoon. 1992. "Case Report: Acute Alcohol Intoxication in a Hedgehog (*Erinaceus europaeus*)." *Kleintierpraxis* 37: 329–332.

Shohat-Ophir, G., K. R. Kaun, R. Azanchi, H. Mohammed, and U. Heberlein. 2012. "Sexual Deprivation Increases Ethanol Intake in *Drosophila*." *Science* 335: 1351–1355.

Starmer, W. T., W. B. Heed, and E. S. Rockwood-Sluss. 1977. "Extension of Longevity in *Drosophila mojavensis* by Environmental Ethanol: Differences between Subraces." *Proceedings of the National Academy of Sciences of the United States of America* 74, no. 1: 387–391.

Tyson, N. deG. 1995. "The Milky Way Bar." *Natural History* 103: 16–18.

Wiens, F., A. Zitzmann, M.-A. Lachance, M. Yegles, et al. 2008. "Chronic Intake of Fermented Floral Nectar by Wild Treeshrews." *Proceedings of the National Academy of Sciences of the United States of America* 105, no. 30: 10426–10431.

Chapter 2. Beer in the Ancient World

There is an extensive literature on the history of beer. Excellent general works include McGovern (2009, 2017), Standage (2005), and Bostwick (2014). The McGovern titles are particularly essential for understanding the wider role of beer in the ancient world and how we know about it. Research at Abu Hureyra is summarized by Moore (2003). Ninkasi's brew is widely cited on the Internet and the *Hymn* was translated by Civil (1991). Katz and Maytag (1991) described the first re-creation of Ninkasi's beer using information in the poem. A dissenting view of Sumerian beer is found in Damerow (2012). See Dineley and Dineley (2000) for Neolithic brewing at Skara Brae; Stika (2011) describes the evidence for brewing in the German Iron Age. See McGovern (2017) for a comprehensive account of the re-creation of ancient beers. For early human inebriation, particularly in Europe, see Guerra-Doce (2015).

Bostwick, W. 2014. *A History of the World According to Beer*. New York: W.W. Norton.

Civil, M. 1991. "Modern Breweries Recreate Ancient Beer." *Oriental Institute News and Notes* 132: 1–2, 4.

Damerow, P. 2012. "Sumerian Beer: The Origins of Brewing Technology in Ancient Mesopotamia." *Cuneiform Digital Library Journal* 2012:002. https://cdli.ucla.edu/files/publications/cdlj2012_002.pdf.

Dineley, Merryn, and Graham Dineley. 2000. "From Grain to Ale: Skara Brae, a Case Study." Pp. 196–200 in A. Ritchie, ed., *Neolithic Orkney in Its European Context*. Cambridge: McDonald Institute.

Guerra-Doce, E. 2015. "The Origins of Inebriation: Archaeological Evidence of the Consumption of Fermented Beverages in Prehistoric Eurasia." *Journal of Archaeological Methods and Theory* 22: 751–782.

Katz, S., and F. Maytag. 1991. "Brewing an Ancient Beer." *Expedition* 44: 24–33.

McGovern, P. E. 2009. *Uncorking the Past: The Quest for Wine, Beer and Other Alcoholic Beverages*. Berkeley: University of California Press.

———. 2017. *Ancient Brews, Rediscovered and Re-Created*. New York: W.W. Norton.

Moore, A. M. T. 2003. "The Abu Hureyra Project: Investigating the Beginning of Farming in Western Asia." Pp. 59–74 in A. J. Ammerman and P. Biagi, eds., *The Widening Harvest. The Neolithic Transition in Europe: Looking Back, Looking Forward*. Boston: Archaeological Institute of America.

Standage, T. 2005. *A History of the World in Six Glasses*. New York: Walker & Co.

Stika, H. P. 2011. "Early Iron Age and Late Mediaeval Malt Finds from Germany—Attempts at Reconstruction of Early Celtic Brewing and the Taste of Celtic Beer. *Archaeological and Anthropological Sciences* 3: 41–48.

Chapter 3. Innovation and an Emerging Industry

William Bostwick (2014) engagingly discusses the history of brewing in Europe and the United States, as does Pete Brown in a series of highly entertaining books (Brown 2003, 2006, 2010, 2012) that range across the world. Nuggets of history (as well as much else) are also to be dug out of such surveys as Alworth (2015) and Bernstein (2013); and, with the usual caveats, extensive information is also to be found on the Internet.

Alworth, J. 2015. *The Beer Bible*. New York: Workman.

Bernstein, J. M. 2013. *The Complete Beer Course*. New York: Sterling Epicure.

Bostwick, W. 2014. *A History of the World According to Beer*. New York: W. W. Norton.

Brown, P. 2003. *Man Walks into a Pub: A Sociable History of Beer*. London: Pan.

———. 2006. *Three Sheets to the Wind. 300 Bars in 13 Countries: One Man's Quest for the Meaning of Beer*. London: Pan.

———. 2010. *Hops and Glory: One Man's Search for the Beer that Built the British Empire*. London: Pan.

————. 2012. *Shakespeare's Pub: A Barstool History of London as Seen through the Windows of Its Oldest Pub — The George Inn.* New York: St. Martin's Griffin.

Chapter 4. Beer-Drinking Cultures

Brown (2006) is as engaging a general account as you will find of beer-drinking cultures around the world. Schivelbusch (1992) describes the camaraderie of the bar. Finch-Hatton's observations of Australian drinking habits are in his autobiography (1886). The *Guardian* report of the Australian High Court's ruling on Queensland's indigenous alcohol laws can be found at https://www .theguardian.com/world/2013/jun/19/australia-indigenous-alcohol-law (accessed June 7, 2018). The numerous guides to Munich's Oktoberfest include Wolff (2013). For the history of innkeeping in Britain, see Brown (2012).

Brown, P. 2006. *Three Sheets to the Wind. 300 Bars in 13 Countries: One Man's Quest for the Meaning of Beer.* London: Pan.

————. 2012. *Shakespeare's Pub: A Barstool History of London as Seen Through the Windows of Its Oldest Pub — The George Inn.* New York: St. Martin's Griffin.

Finch-Hatton, H. 1886. *Advance Australia! An account of eight years' work, wandering, and amusement, in Queensland, New South Wales and Victoria.* London: W. H. Allen.

Schivelbusch, W. 1992. *Tastes of Paradise: A Social History of Spirits, Stimulants, and Intoxicants.* New York: Pantheon.

Wolff, M. 2013. *Meet Me in Munich: A Beer Lover's Guide to Oktoberfest.* New York: Skyhorse.

Chapter 5. Essential Molecules

We have addressed the molecular and chemical backgrounds to alcoholic beverages in greater detail than we have done here in Tattersall and DeSalle (2015). The barley genome was described by the International Barley Genome Sequencing Consortium (2012). Natsume et al. (2014) summarized the draft genome of *Humulus lupulus*. The yeast genome was reported by Mewes et al. (1997), and the sequencing of large numbers of brewer's yeasts was reported by Monerawela and Bond (2017). For the inner workings of the STRUCTURE program see Pritchard (2003) and Earl (2012). Emanuelli et al. (2013) is the source for the grapevine PCA and STRUCTURE figures presented in this chapter.

Earl, D. A. 2012. "STRUCTURE HARVESTER: A Website and Program for Visualizing STRUCTURE Output and Implementing the Evanno Method." *Conservation Genetics Resources* 4 (2): 359–361.

Emanuelli, F., S. Lorenzi, L. Grzeskowiak, V. Catalano, M. Stefanini, M. Troggio, S. Myles, et al. 2013. "Genetic Diversity and Population Structure Assessed by SSR and SNP Markers in a Large Germplasm Collection of Grape." *BMC Plant Biology* 13, no. 1: 39.

International Barley Genome Sequencing Consortium. 2012. "A Physical, Genetic and Functional Sequence Assembly of the Barley Genome." *Nature* 491 (7426): 711–717.

Mewes, H. W., K. Albermann, M. Bähr, D. Frishman, A. Gleissner, J. Hani, K. Heumann, et al. 1997. "Overview of the Yeast Genome." *Nature* 387 (6632): 7–8.

Monerawela, C., and U. Bond. 2017. "Brewing up a Storm: The Genomes of Lager Yeasts and How They Evolved." *Biotechnology Advances* 35: 512–519.

Natsume, S., H. Takagi, A. Shiraishi, J. Murata, H. Toyonaga, J. Patzak, M. Takagi, et al. 2014. "The Draft Genome of Hop (*Humulus lupulus*), an Essence for Brewing." *Plant and Cell Physiology* 56 (3): 428–441.

Pritchard, J. K., W. Wen, and D. Falush. 2003. *Documentation for Structure Software: Version 2.* https://web.stanford.edu/group/pritchardlab/software/readme_structure2.pdf (accessed June 7, 2018).

Tattersall, I., and R. DeSalle. *A Natural History of Wine.* Yale University Press, 2015.

Chapter 6. Water

The Dry Earth Theory and the impact of the Vesta research on it was reported on by Fazekas (2014), and the veracity of Archimedes' "eureka!" is discussed by Biello (2006). References for the French and U.S. water hardness maps can be found online.

Biello, D. 2006. "Fact or Fiction? Archimedes Coined the Term 'Eureka!' in the Bath." *Scientific American,* December 8, 2006.

Fazekas, A. 2014. "Mystery of Earth's Water Origin Solved." *National Geographic,* October 30, 2014.

French water hardness data from Wikimedia Commons: https://commons.wikimedia.org/wiki/File:Duret%C3%A9_de_l%27eau_en_France.svg (accessed June 7, 2018).

U.S. Water Hardness Map. *Fresh Cup Magazine,* July 19, 2016. http://www.freshcup.com/us-water-hardness-map (accessed June 7, 2018).

Water Hardness and Beers: https://www.pinterest.com/pin/443112050818231146 (accessed June 7, 2018).

Chapter 7. Barley

The archaeological remains from Ohalo II relevant to the study of barley are described by Weiss et al. (2004, 2005). The *Global Strategy for the Ex-Situ Conservation and Use of Barley Germ Plasm* can be found at the website listed here. The work of von Bothmer et al. (2003) is the source for much of what we know about diversity of barley strains. The *Hordeum* phylogeny discussed in the text comes from Brassac and Blattner (2015). Pankin and von Korff discuss *Hordeum* and the "domestication syndrome." Mascher et al. (2016) and Russell et al. (2016) describe barley exome sequencing and population genomics; the Mascher et al. paper includes the analysis of ancient barley grains. Pourkheirandish and Komatsuda (2007) discuss the brittle rachis trait. Poets et al. (2015) is the source of the barley PCA overlain on the map of Europe and Asia, as well as of the PCA analysis of *Hordeum* landraces. The review by Robin Allaby (2015) is also referenced here. Jonas and de Koning (2013) summarize the workings of genomics selection and genomic prediction. Schmidt et al. (2016) and Nielsen et al. (2016) provide examples of the use of genomic approaches to improve barley characteristics.

Allaby, R. G. 2015. "Barley Domestication: The End of a Central Dogma?" *Genome Biology* 16, no. 1: 176.

Brassac, J., and F. R. Blattner. 2015. "Species-Level Phylogeny and Polyploid Relationships in *Hordeum* (Poaceae) Inferred by Next-Generation Sequencing and in Silico Cloning of Multiple Nuclear Loci." *Systematic Biology* 64, no. 5: 792–808.

Global Strategy for the Ex-Situ Conservation and Use of Barley Germ Plasm. 2014. https://cdn.croptrust.org/wp/wp-content/uploads/2017/02/Barley _Strategy_FINAL_27Oct08.pdf (accessed June 7, 2018).

Jonas, Elisabeth, and Dirk-Jan de Koning. 2013. "Does Genomic Selection Have a Future in Plant Breeding?" *Trends in Biotechnology* 31: 497–504.

Mascher, M., V. J. Schuenemann, U. Davidovich, N. Marom, A. Himmelbach, S. Hübner, A. Korol, et al. 2016. "Genomic Analysis of 6,000-Year-Old Cultivated Grain Illuminates the Domestication History of Barley." *Nature Genetics* 48, no. 9: 1089–1093.

Nielsen, N. H., A. Jahoor, J. D. Jensen, J. Orabi, F. Cericola, V. Edriss, and J. Jensen. 2016. "Genomic Prediction of Seed Quality Traits Using Advanced Barley Breeding Lines." *PloS One* 11, no. 10: e0164494.

Pankin, A., and M. von Korff. 2017. Co-evolution of Methods and Thoughts in Cereal Domestication Studies: A Tale of Barley (*Hordeum vulgare*)." *Current Opinion in Plant Biology* 36: 15–21.

Poets, A. M., Z. Fang, M. T. Clegg, and P. L. Morrell. 2015. "Barley Land-races Are Characterized by Geographically Heterogeneous Genomic Origins." *Genome Biology* 16, no. 1: 173.

Pourkheirandish, M., and T. Komatsuda. 2007. "The Importance of Barley Genetics and Domestication in a Global Perspective." *Annals of Botany* 100: 999–1008.

Russell, J., M. Mascher, I. K. Dawson, S. Kyriakidis, C. Calixto, F. Freund, M. Bayer, et al. 2016. "Exome Sequencing of Geographically Diverse Barley Landraces and Wild Relatives Gives Insights into Environmental Adaptation." *Nature Genetics* 48, no. 9: 1024–1030.

Schmidt, M., S. Kollers, A. Maasberg-Prelle, J. Großer, B. Schinkel, A. Tome-rius, A. Graner, and V. Korzun. 2016. "Prediction of Malting Quality Traits in Barley Based on Genome-wide Marker Data to Assess the Potential of Genomic Selection." *Theoretical and Applied Genetics* 129, no. 2: 203–213.

von Bothmer, R., T. van Hintum, H. Knüpffer, and K. Sato. 2003. *Diversity in Barley* (Hordeum vulgare), vol. 7. New York: Elsevier Science.

Weiss, E., M. E. Kislev, O. Simchoni, and D. Nadel. 2005. "Small-Grained Wild Grasses as Staple Food at the 23,000-Year-Old Site of Ohalo II, Israel." *Economic Botany* 588: 125–134.

Weiss, E., W. Wetterstrom, D. Nadel, and O. Bar-Yosef. 2004. "The Broad Spectrum Revisited: Evidence from Plant Remains." *Proceedings of the National Academy of Sciences, USA* 101: 9551–9555.

Chapter 8. Yeast

For an examination of the microbial world in and on us, see DeSalle and Perkins (2015), and Dunn (2011). The work by Rytas Vilgalys mentioned in this chapter is summarized in James et al. (2006). The *Saccharomyces* phylogeny is based on work by Cliften et al. (2003), and the life cycle of *Saccharomyces* is from Tsai et al. (2008). The Liti et al. (2009) study of the *cerevisiae* progenitor, and the yeast strain relationships study by the Verstrepen group (Gallone et al. 2016) are also referenced below. The PCA phylogeny and STRUCTURE analyses of the yeast strains are from Gallone et al. (2016). Differences between lager yeasts are discussed by Berlowska, Kregiel, and Rajkowska (2015). The variability of wine yeast strains is discussed in Borneman et al. (2016). The bioreactor approach of Alshakim Nelson is described in the *Economist*, as referenced below.

Berlowska, J., D. Kregiel, and K. Rajkowska. 2015. "Biodiversity of Brewery Yeast Strains and Their Fermentative Activities." *Yeast* 32, no. 1: 289–300.

"A Better Way to Make Drinks and Drugs." *Economist,* July 6, 2017.

Borneman, A. R., A. H. Forgan, R. Kolouchova, J. A. Fraser, and S. A. Schmidt. 2016. "Whole Genome Comparison Reveals High Levels of Inbreeding and Strain Redundancy across the Spectrum of Commercial Wine Strains of *Saccharomyces cerevisiae*." *G3: Genes, Genomes, Genetics* 6, no. 4: 957–971.

Cliften, P., P. Sudarsanam, A. Desikan, L. Fulton, B. Fulton, J. Majors, R. Waterston, B. A. Cohen, and M. Johnston. 2003. "Finding Functional Features in *Saccharomyces* Genomes by Phylogenetic Footprinting." *Science* 301, no. 5629: 71–76.

DeSalle, R., and S. L. Perkins. 2015. *Welcome to the Microbiome: Getting to Know the Trillions of Bacteria and Other Microbes in, on, and around You.* New Haven, CT: Yale University Press.

Dunn, Rob. 2011. *The Wild Life of Our Bodies.* New York: Harper Collins.

Gallone, B., J. Steensels, T. Prahl, L. Soriaga, V. Saels, B. Herrera-Malaver, A. Merlevede, et al. 2016. "Domestication and Divergence of *Saccharomyces cerevisiae* Beer Yeasts." *Cell* 166, no. 6: 1397–1410.

James, T. Y., F. Kauff, C. L. Schoch, P. B. Matheny, V. Hofstetter, C. J. Cox, G. Celio, et al. 2006. "Reconstructing the Early Evolution of Fungi Using a Six-Gene Phylogeny." *Nature* 443 (7113): 818.

Liti, G., D. M. Carter, A. M. Moses, J. Warringer, L. Parts, S. A. James, R. P. Davey, et al. 2009. "Population Genomics of Domestic and Wild Yeasts." *Nature* 458, no. 7236: 337.

Tsai, I. J., D. Bensasson, A. Burt, and V. Koufopanou. 2008. "Population Genomics of the Wild Yeast *Saccharomyces paradoxus:* Quantifying the Life Cycle." *Proceedings of the National Academy of Sciences, USA* 105, no. 12: 4957–4962.

Chapter 9. Hops

The phylogenetic history of the Cannabaceae is published in Yang et al. (2013). The quote about religion and beer in sixteenth-century England is from von Rycken Wilson (1921). Dresel et al. (2016) describe the chemical characteristics of ninety hop strains. HopBase can be accessed at the website listed below.

Dresel, M., C. Vogt, A. Dunkel, and T. Hofmann. 2016. "The Bitter Chemodiversity of Hops (*Humulus lupulus* L.)." *Journal of Agricultural and Food Chemistry* 64, no. 41: 7789–7799.

HopBase. http://hopbase.cgrb.oregonstate.edu (accessed June 7, 2018).

von Rycken Wilson, E. 1921. "Post-Reformation Features of English Drinking." *American Catholic Quarterly* 46: 134–155.

Yang, M.-Q., R. van Velzen, F. T. Bakker, A. Sattarian, D.-Z. Li, and T.-S. Yi. 2013. "Molecular Phylogenetics and Character Evolution of Cannabaceae." *Taxon* 62, no. 3: 473–485.

Chapter 10. Fermentation

These four publications provide a good introduction to fermentation and its applications.

Buchholz, K., and J. Collins. 2013. "The Roots—A Short History of Industrial Microbiology and Biotechnology." *Applied Microbiology and Biotechnology* 97, no. 9: 3747–3762.

Jelinek, B. 1946. "Top and Bottom Fermentation Systems and Their Respective Beer Characteristics." *Journal of the Institute of Brewing* 52, no. 4: 174–181.

Parakhia, M., R. S. Tomar, and B. A. Golakiya. 2015. *Overview of Basics and Types of Fermentation.* Munich: GRIN Publishing.

Thomas, K. 2013. "Beer: How It's Made—The Basics of Brewing." *Liquid Bread: Beer and Brewing in Cross-Cultural Perspective* 7: 35.

Chapter 11. Beer and the Senses

For an overview of the senses, see our book on *The Brain,* published in 2012. Crick quote is from Crick (1990). See also the four references here by Charles Spence, who has been at the forefront of studying sensory impact on taste, and consumer reaction. Schott (1993) examines Penfield's homunculus. Christiaens et al. (2014) describe the origin of fungal aroma genes. Bushdid et al. (2014) is the source for the "trillions of smells" claim. Meilgaard, Carr, and Civille (2006) summarize the lead author's work on beer and taste.

Bushdid, C., M. O. Magnasco, L. B. Vosshall, and A. Keller. 2014. "Humans Can Discriminate More Than 1 Trillion Olfactory Stimuli." *Science* 343, no. 6177: 1370–1372.

Christiaens, J. F., L. M. Franco, T. L. Cools, L. De Meester, J. Michiels, T. Wenseleers, B. A. Hassan, E. Yaksi, and K. J. Verstrepen. 2014. "The Fungal Aroma Gene ATF1 Promotes Dispersal of Yeast Cells through Insect Vectors." *Cell Reports* 9, no. 2: 425–432.

Crick, F. 1990. *Astonishing Hypothesis: The Scientific Search for the Soul.* New York: Scribners.

DeSalle, R., and I. Tattersall. 2012. *The Brain: Big Bangs, Behaviors, and Beliefs.* New Haven, CT: Yale University Press.

Meilgaard, M. C., B. T. Carr, and G. V. Civille. 2006. *Sensory Evaluation Techniques.* Boca Raton, FL: CRC Press.

Schott, Geoffrey D. 1993. "Penfield's Homunculus: A Note on Cerebral Cartography." *Journal of Neurology, Neurosurgery & Psychiatry* 56, no. 4: 329–333.

Spence, C. 2015. "On the Psychological Impact of Food Colour." *Flavour* 4, no. 1: 21.

———. 2016. "Sound—The Forgotten Flavour Sense." *Multisensory Flavor Perception: From Fundamental Neuroscience Through to the Marketplace:* 81.

Spence, C., and G. Van Doorn. 2017. "Does the Shape of the Drinking Receptacle Influence Taste/Flavour Perception? A Review." *Beverages* 3, no. 3: 33.

Spence, C., and Q. J. Wang. 2015. "Sensory Expectations Elicited by the Sounds of Opening the Packaging and Pouring a Beverage." *Flavour* 4, no. 1: 35.

Chapter 12. Beer Bellies

More information on the beer belly can be found in Schütze et al (2009), Shelton and Knott (2014), and Bobak, Skodova, and Marmot (2003). Falony et al. (2016) is the source for our discussion of beer and microbiomes. For a great review of the impact of alcohol on the kidneys see Epstein (1997). Lu and Cederbaum (2008) describe the CYP2E1 gene interaction with alcohol, and the biology of ADH variants is discussed by Mulligan et al. (2003). GWAS and alcoholism is discussed in Bierut et al. (2010).

Bierut, L. J., A. Agrawal, K. K. Bucholz, K. F. Doheny, et al. 2010. "A Genome-Wide Association Study of Alcohol Dependence." *Proceedings of the National Academy of Sciences, USA* 107, no. 11: 5082–5087.

Bobak, M., Z. Skodova, and M. Marmot. 2003. "Beer and Obesity: A Cross-Sectional Study." *European Journal of Clinical Nutrition* 57, no. 10: 1250.

Epstein, M. 1997. "Alcohol's Impact on Kidney Function." *Alcohol Health Research World* 21: 84–92.

Falony, G., M. Joossens, S. Vieira-Silva, J. Wang, Y. Darzi, K. Faust, A. Kurilshikov, et al. 2016. "Population-Level Analysis of Gut Microbiome Variation." *Science* 352 (6285): 560–564.

Lu, Y., and A. I. Cederbaum. 2008. "CYP2E1 and Oxidative Liver Injury by Alcohol." *Free Radicals in Biology and Medicine* 44, no. 5: 723–738.

Mulligan, C., R. W. Robin, M. V. Osier, N. Sambughin, et al. 2003. "Allelic Variation at Alcohol Metabolism Genes (ADH1B, ADH1C, ALDH2) and Alcohol Dependence in an American Indian Population." *Human Genetics* 113, no. 4: 325–336.

Schütze, M., M. Schulz, A. Steffen, et al. 2009. "Beer Consumption and the

'Beer Belly': Scientific Basis or Common Belief?" *European Journal of Clinical Nutrition* 63, no. 9: 1143–1149.

Shelton, N. J., and C. S. Knott. 2014. "Association between Alcohol Calorie Intake and Overweight and Obesity in English Adults." *American Journal of Public Health* 104, no. 4: 629–631.

Chapter 13. Beer and the Brain

For the origin of the term "voluntary madness" and a fairly detailed look at how we become drunk, see our book on *Wine* (Tattersall and DeSalle 2015). For a more general discussion of the origin, structure, and function of the human brain, see our book *The Brain* (DeSalle and Tattersall 2012).

DeSalle, R., and I. Tattersall. 2012. *The Brain: Big Bangs, Behaviors, and Beliefs.* New Haven, CT: Yale University Press.

Tattersall, I., and R. DeSalle. *A Natural History of Wine.* New Haven, CT: Yale University Press, 2015.

Chapter 14. Beer Phylogeny

These eight websites, last accessed in June of 2018, offer evolutionary trees or other representations of beer taxonomy. For a primer of phylogenetics methods, see DeSalle and Rosenfeld (2013). The Beer Judge Certification Program guidelines can be found at the BJCP website. The beer periodic table and the 33beers scoring booklets with taste wheels can similarly be found at their websites.

https://www.popchartlab.com.

http://www.allposters.com.

https://cratestyle.com.

http://phylonetworks.blogspot.com/2015/11/are-taxonomies-networks.html.

https://commons.wikimedia.org/wiki/File: Beer types diagram.svg.

http://randomrow.com/phylogeny-of-beer.

https://twitter.com/dangraur/status/642028902982901760.

http://clydesparks.com/everything-you-need-to-know-about-beer-in-one
    -chart-infographic.

Beer Judge Certification Program (BJCP): https://www.bjcp.org/docs/2015
    _Guidelines_Beer.pdf.

Beer Periodic Table: https://www.posterazzi.com.

DeSalle, Rob, and Jeffrey Rosenfeld. 2013. *Phylogenomics: A Primer.* New York: Garland Science.

33beers scoring booklets: http://33books.com.

Chapter 15. The Resurrection Men

The key reference to the making of ancient beers in both early and modern times is McGovern (2017). The first re-creation of a Sumerian beer was recounted by Katz and Maytag (1991), who referred to the *Hymn to Ninkasi* as translated and quoted by Civil (1991). Samuel (1996a, b) describes the analysis of the Amarna brewery evidence. Calagione (2011) outlines the role of the Dogfish Head Brewery in replicating ancient beers, while Brown (2012) describes his adventures re-creating an authentic IPA.

Brown, P. 2012. *Hops and Glory: One Man's Search for the Beer That Built the British Empire*. London: Pan.

Calagione, S. 2011. *Brewing up a Business*. Revised and updated edition. Hoboken, NJ: Wiley.

Civil, M. 1991. "Modern Breweries Recreate Ancient Beer." *Oriental Institute News and Notes* 132: 1–2, 4.

Katz, S., and F. Maytag. 1991. "Brewing an Ancient Beer." *Expedition* 44: 24–33.

McGovern, P. E. 2017. *Ancient Brews, Rediscovered and Re-Created*. New York: W. W. Norton.

Samuel, D. 1996a. "Investigation of Ancient Egyptian Baking and Brewing Methods by Correlative Microscopy." *Science* 273: 488–490.

———. 1996b. "Archaeology of Ancient Egyptian Beer." *Journal of the American Society of Brewing Chemists* 54: 3–12.

Chapter 16. The Future of Brewing

Bostwick (2014) provides an engaging survey of the history of beer in America, and Brown (2012) does a similar service for the United Kingdom. Brown (2006, 2012) is also amusingly articulate on the results of the globalization of brewing. Extensive information on CAMRA may be found on its website: http://www.camra.org.uk (accessed June 7, 2018). Excellent accounts of the rise of the craft beer movement in the United States may be found in Accitelli (2013) and Hindy (2014); for a general appraisal see Elzinga, Tremblay, and Tremblay (2015). This is a fast-moving field on which the latest information is best found by trawling (cautiously) through the Internet.

Accitelli, T. 2013. *The Audacity of Hops: The History of America's Craft Beer Revolution*. Chicago: Chicago Review Press.

Bostwick, W. 2014. *A History of the World According to Beer*. New York: W. W. Norton.

Brown, P. 2006. *Three Sheets to the Wind. 300 Bars in 13 Countries: One Man's Quest for the Meaning of Beer*. London: Pan.

———. 2012. *Shakespeare's Pub: A Barstool History of London as Seen through the Windows of Its Oldest Pub — The George Inn*. New York: St. Martin's Griffin.

Elzinga, K. G., C. H. Tremblay, and V. J. Tremblay. 2015. "Craft Beer in the United States: History, Numbers, and Geography." *Journal of Wine Economics* 10: 242–274.

Hindy, S. 2014. *The Craft Beer Revolution: How a Band of Microbrewers Is Transforming the World's Favorite Drink*. New York: Palgrave Macmillan.

Index

Note: Page numbers in italics indicate illustrations.